ALKALI-ION BATTERIES

Edited by **Dongfang Yang**

Alkali-ion Batteries
http://dx.doi.org/10.5772/61429
Edited by Dongfang Yang

Contributors

Lianfeng Duan, Feifei Zhang, Limin Wang, Rabin Bissessur, Douglas Dahn, Evans A. Monyoncho, Victoria Trenton, Verónica Palomares, Paula Serras, Teofilo Rojo, Natasha Ross, Xiaowei Teng, Daniel Charles, Yunhua Yu, Yuan Liu, Xiaoping Yang, Huiling Duan, Yingjie Liu, Takayuki Ichikawa, Liang Zeng, Koji Kawahito, Mitsunobu Sato, Hiroki Nagai

Published by InTech

Janeza Trdine 9, 51000 Rijeka, Croatia

Publishing Process Manager Iva Lipovic

Technical Editor InTech DTP team

Cover InTech Design team

First published May, 2016
Printed in Croatia

Additional hard copies can be obtained from orders@intechopen.com

Alkali-ion Batteries, Edited by Dongfang Yang

p. cm.

Print ISBN 978-953-51-2395-8
Online ISBN 978-953-51-2396-5

For EU product safety concerns:
IN TECH d.o.o., Prolaz Marije Krucifikse Kozulić 3, 51000 Rijeka, Croatia,
info@intechopen.com or visit our website at intechopen.com.

Meet the editor

Dongfang Yang received his Ph.D. in Physical Chemistry from the University of Guelph in 1995. He joined the National Research Council of Canada in London, Ontario, in 2001 and is now a Senior Research Officer. His current research interests include laser materials processing; pulsed laser, sputtering and e-beam deposition of thin films; new materials development for energy storage devices; chemical and optical sensors development and electrochemical studies of organic adsorption and self-assembly monolayer. He is currently serving as a member of editorial board for five scientific journals including The Research & Reviews in Electrochemistry, Current Chemical Research and journal Sensors & Transducers. He also holds an Adjunct Professorship at both Lakehead University and Western University in Ontario, Canada.

Contents

Preface

This book covers selected topics in different aspects of science and technology of alkali-ion batteries written by experts from the international scientific community. Through these 9 chapters, the reader will have access to the most recent research and development findings on alkali-ion batteries through original research studies and literature reviews. This book covers inter-disciplinary aspects of alkali-ion batteries, including new progresses made in material chemistry, micro/nano structural designs, computational and theoretical models and understanding of structural changes during electrochemical processes of alkali-ion batteries. Chapter 1 provides an excellent review of the recent progresses made in lithium-ion battery anode materials based on electrospun carbon nanofibers and their composites. Chapter 2 presents a new method to improve the electrochemical performance of $LiMn_2O_4$ cathode active materials by co-precipitation with alloys of Au-Fe, Au-Pd and Au-Pt. Chapter 3 emphasizes on novel micro/nano structural designs of the cathode using composites of core-shell hierarchical porous carbon spheres, graphene, graphene oxides, polymers and their hybrids with sulfur for lithium sulfur batteries. These composite sulfur cathodes can effectively inhibit the dissolution and migration of polysulfides into the electrolyte and accommodate the volume expansion during discharge, therefore improve the cyclic durability and performance of lithium sulfur batteries. Chapter 4 explores work on exploiting metal hydride (MH)-based materials as potential negative materials for lithium-ion batteries, while Chapter 5 exploits a new solid electrolyte or electrode active materials for Li-ion battery by intercalating POEGO into V_2O_5•nH2O. Chapter 6 reveals an innovative thin-film electrochromic/Li-ion battery device and a thin-film Li-ion battery device fabricated on conductive glass substrates that can be charged using solar light irradiation. Chapter 7 presents a theoretic model to predict the concentration and stress fields inside particles of electrode active materials of Li-ion battery during discharge/charge process. Chapter 8 gives an excellent literature review on recent developments on sodium-vanadium fluorophosphates for Na-ion batteries with good electrochemical performance, e.g., high-specific capacity at moderate/high cycling rates, and long cycle life, and has great potential as cathode materials for Na-ion batteries. Chapter 9 presents excellent works on understanding the structure changes during electrochemical processes of ionic transport between the layers of nanostructural vanadium oxide materials. Using multiple spectroscopic tools, such as synchrotron X-ray pair distribution function, in situ X-ray diffraction, as well as SEM, TEM and electrochemical measurement, the authors of the chapter brings convincing insight of ions transportation between the layers of V_2O_5 nanostructure.

This publication provides a snapshot of current efforts to improve alkali-ion battery performance and will be useful for the researchers and students who work in the area of electrochemical energy storage. I gratefully acknowledge all chapter authors for their enthusiastic and collaborative contributions and thank Ms. Iva Lipović, Publishing Process Manager, for her guidance and support in preparing this book.

Dongfang Yang, Ph.D.
Senior Research Officer
Automotive and Surface Transportation
National Research Council Canada
London, Ontario, Canada

Section 1

Li-ion Battery

Chapter 1

Carbon Nanofiber-Based Materials as Anode Materials for Lithium-Ion Batteries

Yunhua Yu, Yuan Liu and Xiaoping Yang

Additional information is available at the end of the chapter

http://dx.doi.org/10.5772/63235

Abstract

Considerable efforts have been devoted to the research of high-performance and long-lifespan lithium-ion batteries (LIBs) for their applications in large-scale power units. As one of the most important components in LIBs, anode plays an important role in determining the overall performance of LIBs. Nowadays, graphite has been the most successfully commercialized anode material. However, its limited theoretical capacity (372 mA h g^{-1}) and limited power density seems insufficient for the next-generation LIBs. To overcome these problems, new materials with fundamentally higher capacity and higher power density are urgently needed. Recently, there is an ever-increasing interest in developing novel carbonaceous nanomaterials to replace graphite as the anode materials for LIBs. Such materials have included carbon spheres, carbon nanotubes, carbon nanofibers (CNFs), porous monoliths, and graphene. Among these alternative forms of carbon, CNFs and its morphological-controlled derivatives (such as porous or hollow CNFs) have attracted much attention due to their unique and interesting properties such as one-dimensional (1D) nanostructure, good electronic conductivity, and large surface areas. Moreover, these CNFs can be used to encapsulate various second phases to form some functional composite, meeting further requirements including higher energy density, higher power density or flexible requirements, for the advanced LIB operation.

Electrospinning is considered as a simple, versatile, and cost-effective industry-viable technology for preparing various CNFs and their composites in a continuous process, with controllable morphology. Therefore, in this chapter, we have summarized some recent progresses in electrospun nanofibrous carbon-based anode for LIBs, covering the structure evolution from solid CNFs into morphology-designed CNFs, and then their composites with various functional nanoparticles. We anticipate this paper can offer some useful information for some researchers in the area of energy storage and conversion and can inspire them.

Keywords: electrospinning, carbon-based nanofibers, anode, lithium-ion batteries, nanomaterials

1. Introduction

Energy conversion and storage become more and more important in the context of the increasing global energy demand because of the inadequacy of fossil fuels, climate change, and deteriorating environmental conditions. Currently, among the available energy conversion and storage technologies, lithium-ion battery (LIB) is the most versatile and successful technology that possess high-energy densities (2–3 times higher than conventional batteries), no memory effects, relatively slow self-discharge rates, and longer battery lifetimes, and therefore they have received intense attention from both the academic community and industry as the dominant power source in hybrid electric vehicles (HEVs), plug-in hybrid electric vehicles (PHEVs), and full electric vehicles (EVs) [1–4]. For further enhancing the performance of LIBs, many studies concentrated on changing either the chemical composition or macroscopic structure of the components [5, 6].

As one of the most important components in LIBs, anode plays an important role in determining the overall performance of LIBs. At present, most commonly used anode materials for commercial LIBs are graphite powders that have limited theoretical capacity (372 mA h g^{-1}) and long diffusion pathways for the lithium ions [7]. This may result in low energy and low power densities, which cannot meet the ever-expanding demands for next-generation LIBs. To resolve the problem, a variety of nanostructured carbonaceous materials have been investigated as anode materials for LIBs, such as carbon nanobeads [8], hollow carbon nanospheres [9, 10], carbon nanotubes [11–13], carbon nanofibers (CNFs) [14–16], graphenes [17–19], and their composites [20–22].

Among various carbon nanostructures, CNFs and its morphological-controlled derivatives (such as porous or hollow CNFs) have attracted much attention because they could provide an enhanced surface-to-volume ratio for the electrode–electrolyte interface, short transport lengths for ionic transport, and efficient one-dimensional (1D) electron transport along the longitudinal direction when compared to the powder materials [23]. Moreover, these CNFs can be used to encapsulate various second phases to form functional composite, meeting the ever-growing demand for advanced batteries. Electrospinning has been widely used as a simple, versatile, and cost-effective industry-viable technology to prepare various CNFs and their composites in a continuous process, with controllable morphology and compositions [24–42]. The principle of electrospinning has been well introduced in several excellent reviews on electrospun materials for energy-related applications [43–46]. In this chapter, we have summarized some recent advances in the area of 1D CNF-based materials for LIB anodes, covering the structure evolution from electrospun solid CNFs into morphology-constructed porous CNFs, and their composites with various functional nanoparticles.

2. Electrospun CNFs

2.1. Solid CNFs

Carbon materials including graphite, graphene, fullerenes, carbon nanotubes, and CNFs have attracted tremendous attention in both fundamental research and industrial applications, especially in the applications of energy storage and conversion devices such as LIBs [47, 48]. Among these various carbon materials, 1D electrospinning-derived CNFs are of high interest as potential anode materials due to their high-specific surface area, good conductivity, and structural stability, which are the key factors influencing the electrochemical properties of carbon electrodes [23, 49–52]. The 1D nature of the CNF anode not only facilitates the electron transport along the axial direction, but also reduces the lithium-ion diffusion distance through short radial direction, both of which are beneficial for the improved specific capacity and rate capability.

The CNFs can be derived from many synthetic or natural polymeric precursors such as polyacrylonitrile (PAN), polyvinyl pyrrolidone (PVP), polyvinyl alcohol (PVA), lignin, or cellulose. It should be noted that the physical and chemical properties of CNFs highly depend on the chemical structure of the carbon precursors. Currently, the commonly used precursor for CNFs is PAN, which has good spinnability, and can yield a high amount of carbon residue after simple stabilization and carbonization processes [44]. Moreover, owing to the robust integrated network structure and good electrical conductivity, PAN-derived CNF webs can be directly used as the anode materials without adding any adhesive and conductive additives, which can reduce the weight of anode, and thus improving the energy density of a full cell [53]. Endo's group [23] synthesized the PAN-derived CNF webs by the electrospinning technique combined with two-step heat-treatments and investigated their lithium-storage properties along the variation of carbonization temperature (from 700°C to 2800°C). The composition ratio of amorphous carbon and graphitic carbon in these CNFs was demonstrated to affect the reversible capacity, slope or plateau charge–discharge characteristic, and rate/cycling performance. The high-purity CNF web thermally treated at 1000°C shows the high-rate capability (350 mA h g^{-1} at a charge current of 100 mA g^{-1}) owing to the interlinked nanofibers, a large accessible surface area, and relatively good electrical conductivity, which make it an ideal candidate for the anode material of high-power LIBs. Nevertheless, the large-scale applications of PAN-derived CNFs might be hampered by the following two reasons. First, PAN is a relatively expensive synthetic polymer, of which the price varies with that of the crude oil [54]. Second, PAN is hard to dissolve in many solvents, and its most commonly used solvent, dimethylformamide (DMF), is known to be harmful for human beings during the electrospinning process [55].

Water or ethanol/polymer system is a better choice to avoid the aforementioned problems and lower the production cost. PVP is the type of water-soluble polymer, and has been widely used in industry due to its merits of low cost, nontoxicity, and good compatibility with metallic precursors. The preparation process of well-controlled PVP-derived CNFs has also been comprehensively investigated; however, the lithium-storage of this fiber-based electrode is temporarily lower than that of PAN-derived CNFs [54]. Other common water-soluble CNF precursors such as PVA [55, 56], lignin, and cellulose [57, 58] have been used to prepare fibrous electrode; nevertheless, all of them show low mechanical properties as compared with PAN-derived CNFs.

Coaxial electrospinning or coelectrospinning, a breakthrough in the electrospinning method, has been used to prepare core–shell soft-hard CNFs, in which a spinneret consisting of two coaxial capillaries is used, with PAN/DMF as the external solution and mineral oil as the inner solution (**Figure 1**) [59]. After the stabilization and carbonization processes, the soft–hard core–shell CNFs were obtained with shell PAN converted to hard carbon and core mineral oil decomposed to soft carbon. The coaxial CNFs combine the advantages of both hard carbon (possess a high capacity of 400–500 mA h g^{-1}, but poor capacity retention performance) and soft carbon (has a lower, but reversible capacity of 200–300 mA h g^{-1}, however, it shows a very serious voltage hysteresis during the delithiation process), and therefore exhibits enhanced reversible capacity as an anode in LIBs (390 mA h g^{-1} at a charge current of 100 mA g^{-1}) even though the kinetics of the charge process requires further improvement.

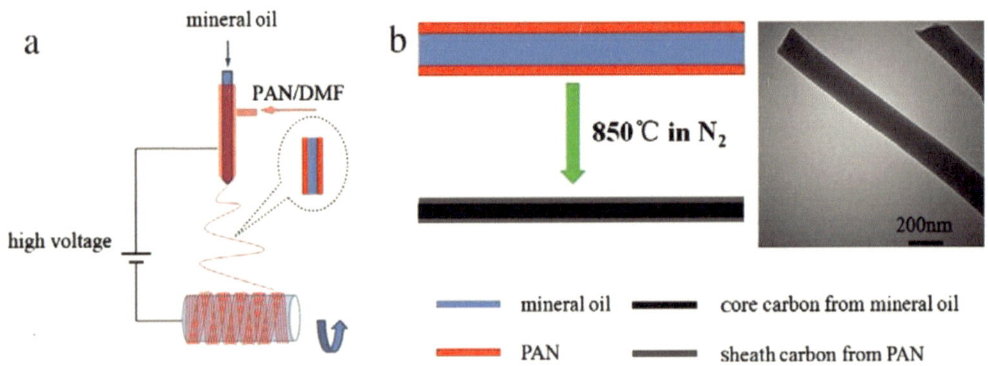

Figure 1. Schematic illustrations of (a) the coaxial electrospinning apparatus and (b) preparation of coaxial CNFs [59].

2.2. Porous CNFs

Recent research has showed that the introduction of various porous structures into CNFs could greatly enhance both the specific capacity/capacitance and the rate capability. This is because the incorporated pore can possibly create high-specific surface area that provides more charge transfer. Up to date, many strategies have been used to control the porous structure in CNFs for LIB application. Template-based processes are of great interest for the preparation of porous CNFs with high surface area. Kang's group [60] synthesized porous CNFs by etching off the silica template in CNFs from pyrolysis of the electrospun polyamic acid/tetraethoxysilane (TEOS) nanofibers. The porous CNF electrode showed a high reversible capacity of 445 mA h g^{-1} after 50 cycles, which is higher than that of commercial graphite (372 mA h g^{-1}). The nitrogen adsorption–desorption isotherms showed that the specific surface area can reach to 950 m^2 g^{-1}, which is contributed to the large amounts of micropores. Here, the micropores can serve as the active "reservoir" for absorbing more lithium during cycling, thus improving the

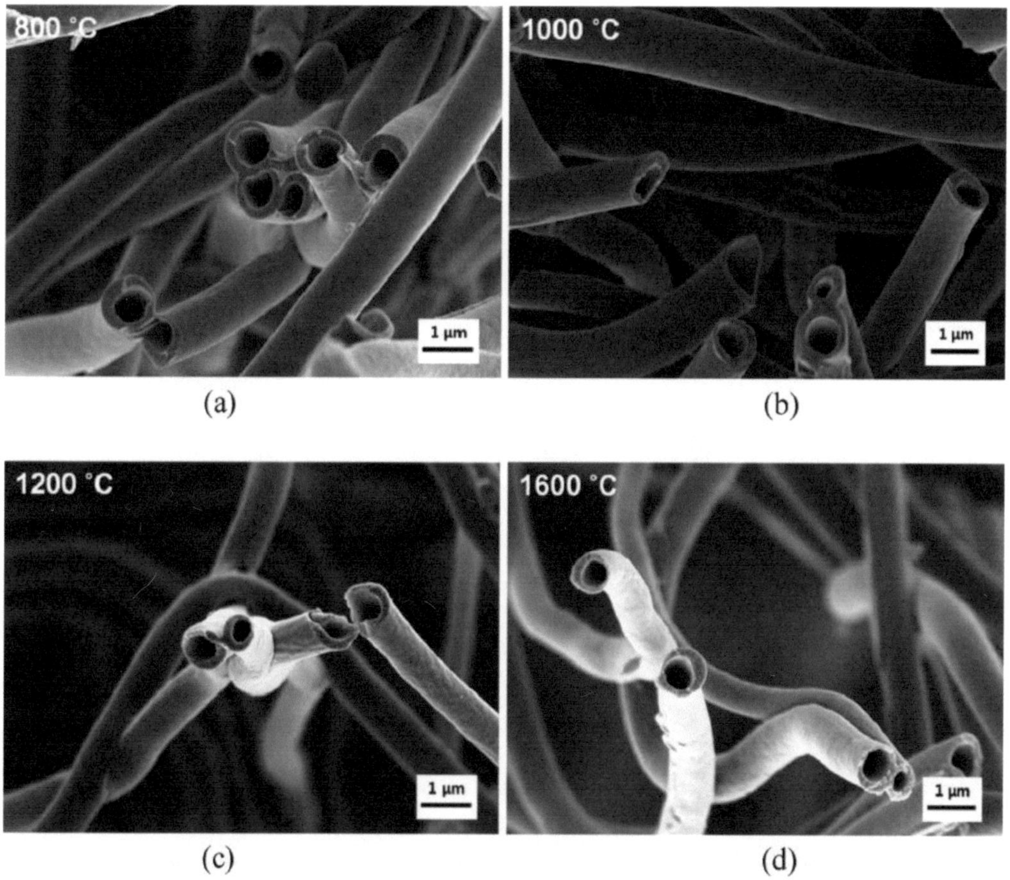

Figure 2. FE-SEM images of HCNFs carbonized at (a) 800°C, (b) 1000°C, (c) 1200°C, and (d) 1600°C [61].

lithium-storage capacity based on the bare CNFs. Lee et al. [61] have utilized coaxial electrospinning to fabricate hollow CNFs (HCNFs) as anode materials and studied the effect of carbonization temperature on the electrochemical performance. Styrene-co-acrylonitrile (SAN) and PAN in DMF solutions were served as the core and shell materials. The as-spun nanofibers were stabilized at 270–300°C for 1 h in air, and then carbonized at 800, 1000, 1200, and 1600°C for 1 h in nitrogen, respectively. During thermal treatment, the linear PAN molecules were transformed to the ladder structure and got carbonized in the following process; meanwhile, the core component burned out leading to the hollow structure (**Figure 2**). The large continuous hollow pore can facilitate the Li^+-carrying electrolyte penetrate into the inner part of CNFs, thus highly reducing the Li^+-diffusion distance, and making the full use of the active lithium-storage part at high charge–discharge rate. The capacities after 10 cycles at a current of 50 mA g^{-1} were 390, 334, 273, and 243 mA h g^{-1} in accordance to their carbonization temperature (800–1600°C), with a very high coulombic efficiency. The reversible discharge capacities are slightly reduced even though four times higher current density is supplied (**Figure 3**).

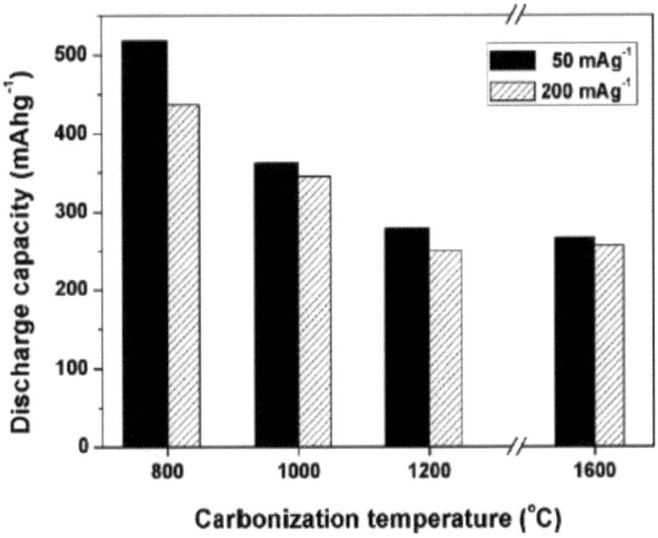

Figure 3. Discharge capacity vs. discharge rate for the HCNFs carbonized at various temperatures [61].

Recently, nitrogen-doped carbon materials are a researchers owing to the high capacity and rate capability [10, 62–65]. In a recent research, Liu et al. fabricated a new type of nitrogen-doped carbon tube by pyrolyzing polydopamine (PDA) using silica nanofibers as templates (**Figure 4**) [65]. The SiO_2 NFs were first fabricated by an electrospinning technique and subsequent calcination in air, and then immersed in a dopamine aqueous solution (pH: ~8.5). Subsequently, the dopamine monomers were covalently joined via aryl–aryl linkages owing to the oxidization and cyclization reactions, forming a PDA coating layer on the surface of the SiO_2 NFs. Then, the core–shell SiO_2 NFs/PDA nanofibers were carbonized at 750°C for 3 h in a N_2 atmosphere. Finally, N-CTs were obtained by etching off the silicate template with sodium hydroxide solution. The N-CTs show a fibrous morphology (diameter, 200–400 nm; length, several micrometers), a typical hollow feature (wall thickness, ~16 nm), and discontinuous and randomly constructed graphene-like layers (the d_{002} interlayer spacing, 0.354 nm) (**Figure 5**). The PDA-derived carbon tubes (N-CTs) as anode materials for LIBs show a remarkable self-improved capacity along cycling. This is contributed to the continuous interlamellar spacing expansion between the graphene-like carbon layers during cycling. (**Figure 6**) Moreover, owing to the unique hollow structure, ultrafine carbon-tube wall, and nitrogen doping, the N-CT electrode shows very high specific capacity, outstanding rate capability, and robust durability, giving a superior reversible capacity of 1635 mA h g^{-1} at 100 mA g^{-1} after 300 cycles and 1103 mA h g^{-1} at 500 mA g^{-1} after 500 cycles. The excellent electrochemical performance makes the N-CTs a potential anode material for the next-generation LIBs.

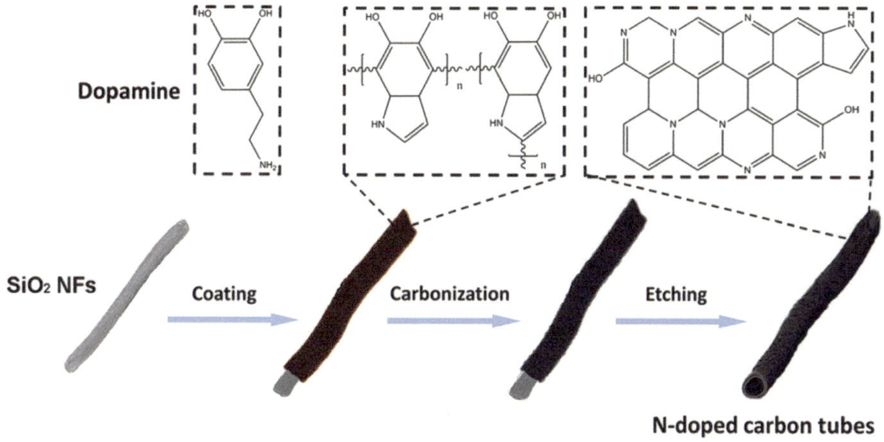

Figure 4. Schematic illustration of the synthesis of N-doped carbon tubes [65].

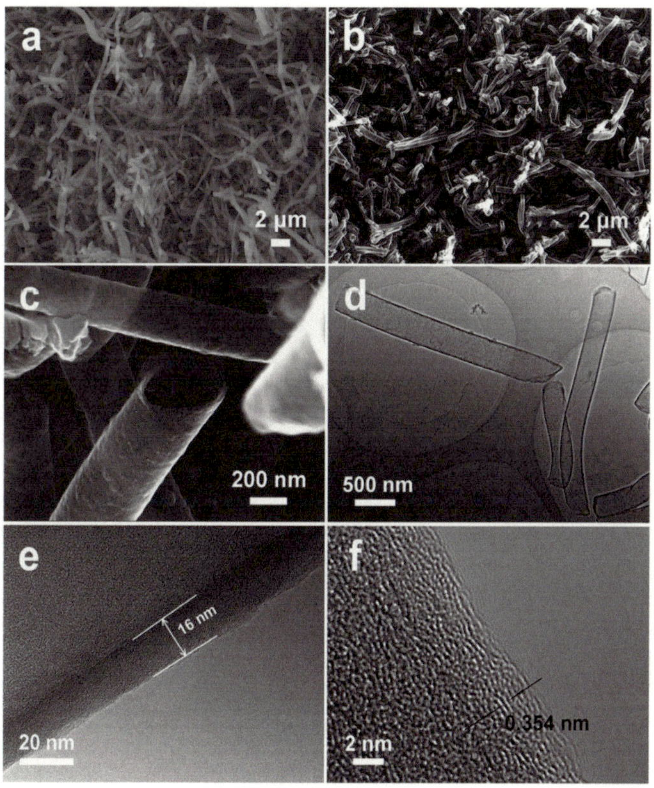

Figure 5. (a) SEM image of SiO_2 NFs, (b and c) SEM images of N-CTs, and (d–f) HR-TEM images of N-CTs [65].

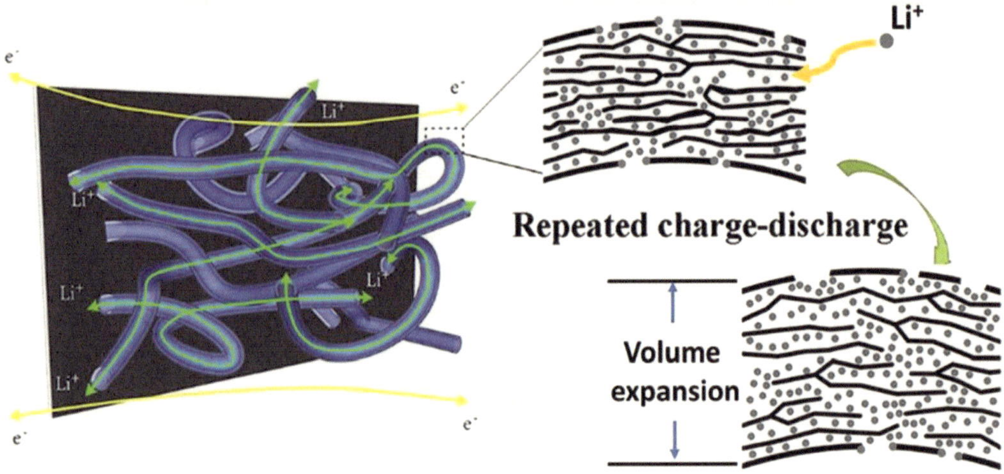

Figure 6. Illustration of lithium-ion storage/transport in N-CTs during the repeated lithiation and delithiation processes [65].

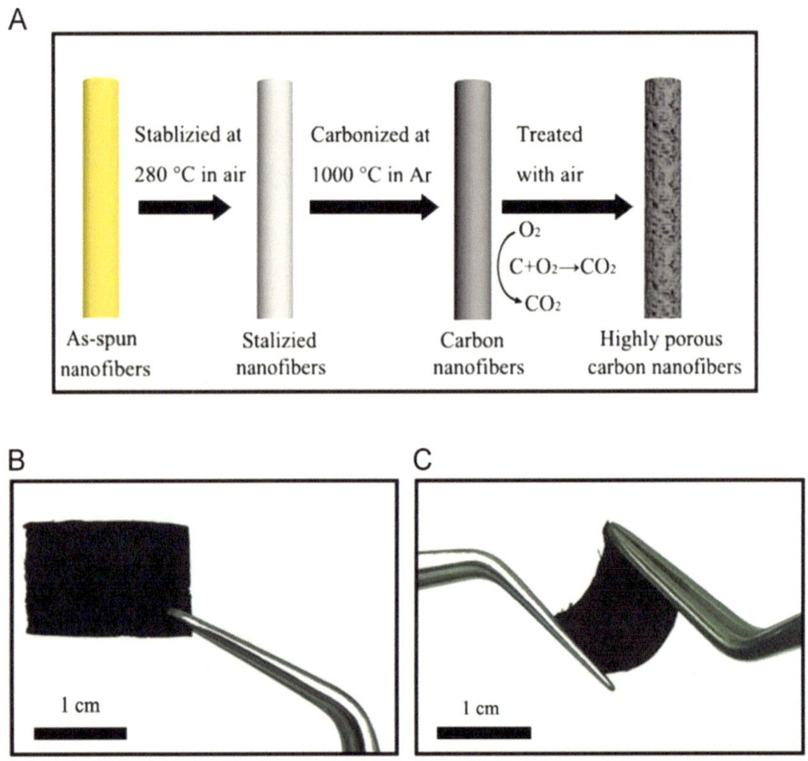

Figure 7. (a) Schematic illustration of the preparation of the HPCNF electrode. (b and c) Photographs of supported and flexible HPCNF film [68].

Besides, some other additives such as $ZnCl_2$ and H_3PO_4 usually act as the activating agents to produce porous structures in CNFs [66, 67]. These activating agents lead to large amounts of micropores in the surface of CNFs, thus providing more surface active sites for absorbing the Li^+. Lately, a novel and simple method (air activation method) have attracted much attention as the method needs no template and activating agents. Yu's group [68] prepared highly porous CNFs (HPCNFs) by two-step carbonization of electrospun PAN nanofibers. During carbonization process at 1000°C in Ar, a certain volume of air were mixed into the Ar flow, where the CNFs were partially burnt off and numerous micro/mesopores were formed simultaneously (**Figure 7**). The as-synthesized HPCNFs exhibit a paper-like external morphology and highly porous internal nanostructure. When used as a binder-free anode in LIBs, the HPCNFs deliver a very high capacity of 1780 mA h g^{-1} at 50 mA g^{-1} after 40 cycles, greatly improved rate capacity and ultralong cycle life (1550 mA h g^{-1} at 500 mA g^{-1} after 600 cycles) in comparison with CNFs. The outstanding electrochemical performance is contributed to the electrospinning-derived 3D porous interconnected networks and the air-activated mesoporous structure in the CNFs that can facilitate the electrolyte into the electrode, thus reducing the Li^+ diffusion distance. Consideration of the low-cost and efficient preparation, this method is hoped to design highly porous materials in large-scale production used for advanced energy-storage devices.

3. Composite anodes with CNFs

The capacities of pure CNFs are insufficient for high-performance batteries. Therefore, various components such as silicon, tin and tin oxides, titanium oxides, and other metal oxide nanoparticles have been loaded into CNFs via an electrospinning process to enhance the performance.

3.1. Si-loaded CNF composite anodes

Silicon nanoparticles (with a theoretical specific capacity of 4200 mA h g^{-1}) could be incorporated into the CNF matrix by electrospinning PAN–Si nanoparticles and the subsequence carbonization to improve the poor cycling performance resulted from large volume changes (~400%) and nanoparticle aggregation upon the alloying and dealloying reaction with Li+ [69, 70]. By optimizing the Si content, Si particles were dispersed homogeneously along the fibers, thus inhibiting the agglomeration of Si nanoparticles and suppressing mechanical failure during Li+ insertion and extraction [71]. Additionally, introduction of various porous structures into CNFs could greatly enhance the specific capacity and rate performance of Si/CNF composite electrodes. For example, porous Si/CNF composites used without binding and conductive additives showed high discharge capacity of 1100 mA h g^{-1} at a high current density of 200 mA g^{-1} [72].

Another example, Si–CNF core–shell fibers with void space in the core section were fabricated by coaxial electrospinning, in which Si–PMMA was chosen as the core and PAN as the shell [73]. After carbonization process, PAN can still remain stable in the shell, while PMMA could be removed to form the void space in the core of the fibers, which can accommodate the volume expansion of Si (**Figure 8**). This unique Si–CNF core–shell structure shows a high capacity of 1384 mA h g^{-1} at a rate of C/10 and an outstanding cycle life of 300 cycles with 99% capacity retention. Another promising strategy to improve the performance of Si/CNFs is adding conductive component such as graphitized carbon [74] or TiO_2 [75] into the electrospun Si–CNF composite. The electrical conduction of the surrounding material significantly improved the reversible capacity and cycling stability.

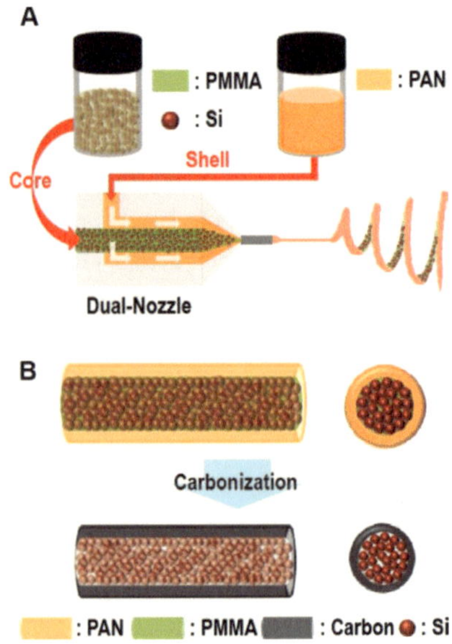

Figure 8. Schematic illustration of Si–CNF core–shell fibers [73].

3.2. Tin-based composite anodes with CNFs

Tin and its oxides have much higher theoretical capacities (Sn: 992 mA h g^{-1}, SnO_2: 780 mA h g^{-1}) than the commercial graphite (372 mA h g^{-1}), but they also suffer from large volume changes and nanoparticle aggregation during cycling, resulting in capacity and stability losses. Dispersing these metallic nanoparticles into CNFs via electrospinning is an efficient approach to overcome these drawbacks because CNFs can hinder particle aggregation, provide continuous long-distance electron transport pathway, support numerous active sites for charge-transfer reactions, and eliminate the need for binding or conducting additive [23]. Yu et al. [53] fabricated a reticular Sn/CNF webs used as anodes for rechargeable LIBs via electrospinning technique and carbonization treatment, and studied the carbonization temperature effect on electrochemical performance of the Sn/CNF webs. It is demonstrated that carbonization temperature will influence Sn grain size, surface area or fiber diameter, and the electrical conductivity of CNFs, which dominate the electrochemical performance of the electrode. The Sn/CNF webs carbonized at 850°C exhibited a reversible capacity of 450 mA h g^{-1} after 30 cycles at a current of 25 mA g^{-1}. Herein, the overall capacity looks low because the direct electro-

http://dx.doi.org/10.5772/63235

spinning technique limits the loading amount of active materials (with ~22 wt% Sn particles), which could be improved by constructing novel nanostructures.

Porous or hollow structure is introduced into Sn/CNF systems to enhance the cycling stabilities and rate capabilities. Sn nanoparticles have been encapsulated into porous multichannel carbon microtubes (SPMCTs) [76] and bamboo-like hollow CNFs (SBCNFs) [77] using a single-nozzle and a coaxial electrospinning technique, respectively. Such porous or hollow carbon shells could provide appropriate void volume to buffer the large volume change, prevent pulverization of the Sn nanoparticles, serve as an electron supplier, and allow more Li^+ access. As a result, both of them showed good cycling stabilities and excellent rate capabilities. Specifically, the SBCNFs display a better cycling stability and a more excellent rate capability with a reversible discharge capacity as high as 480 mA h g^{-1} at 5 C after 100 cycles.

SnO_x/CNF composites have been synthesized by electrospinning and subsequent thermal treatment [78, 79]. For example, ultrauniform SnO_x/carbon nanohybrid (denoted as U-SnO_x/C) has been fabricated by solvent replacement and subsequent electrospinning homogeneous dispersion of SnO_2 nanoparticles in PAN/DMF solution [79]. The strong interaction between SnO_x and nitrogen-containing CNFs (Sn–N bonding) could effectively confine the uniformly embedded SnO_x. This unique nanostructure can not only suppress the agglomeration of SnO_x and tolerate the substantial volume change during cycling, but also enhance the transport of both electrons and ions due to shortened conducting and diffusion pathways. As a consequence, the U-SnO_x/C nanohybrids exhibit a high reversible capacity of 608 mA h g^{-1} after 200 cycles, with excellent rate capability. However, the effect of homogeneous distribution is also limited. Sn- or SnO_2/CNF composites have also been doped with various transition metals, such as Co, Cu and Ni, to improve the cycling stability and rate performance. Co–Sn alloy particles embedded in CNFs improved the cycling stability by increasing the conductivity of the CNF and also enhanced the specific anodic capacity because of different Co–Sn alloys in the structure, as controlled by the carbonization temperature [80]. Incorporation of amorphous Cu into Sn/CNF achieved the highest cycling stability of 490 mA h g^{-1} after 600 cycles at a current density of 156 mA g^{-1} [81]. Addition of Ni into SnO_2 CNF suppressed the reduction of SnO_2 to Sn during carbonization and the agglomeration of SnO_2, thus enhancing the cycling stability [82].

In recent literature, Yu et al. [83–86] have achieved the in-site addition of transition metallic (Ti and Cu) and nonmetallic elements (P and B) into SnO_x/CNF composites for the enhancement of cycling stability and rate performance via the electrospinning technique and subsequent thermal treatments. It was demonstrated that the doped SnO_x nanoparticles were all ultrafine and uniformly dispersed in the conductive CNF matrix, and the doping content should be kept to an optimal value. The incorporation of heteroatoms into SnO_x/CNFs endowed them with the enhancement of cyclic capacity retention and rate performance compared with the pristine SnO_x/CNFs (Table 1) due to the more complete reversible conversion reaction and the higher Li^+-diffusion coefficient. Especially, the addition of Cu into SnO_x/CNFs exists in the form of Cu_2O, which can be transformed into Cu nanoparticles dispersed in a lithia matrix (Li_2O), inhibiting the aggregation of Sn particles in the following alloying–dealloying cycling (**Figure 9**). Meanwhile, the existence of Cu nanoparticles not only improves the Li^+-transport capability and the electronic conductivity of the overall electrode, but also enhances the chemical reacting activity of Sn back to SnO_x during the Li^+-extraction process; therefore, the addition of Cu^+ can endow the SnO_x/CNF electrode with greatly enhanced reversible capacity and rate capability [86].

Materials	Performance	Rates	References
SnO_x/CNFs	640 mA h g^{-1} after 60 cycles	200 mA g^{-1}	85
	230 mA h g^{-1} after 100 cycles	2 A g^{-1}	
Ti-doped SnO_x/CNFs	670.7 mA h g^{-1} after 60 cycles	200 mA g^{-1}	84
	302.1 mA h g^{-1} after 80 cycles	2 A g^{-1}	
P-doped SnO_x/CNFs	676 mA h g^{-1} after 100 cycles	200 mA g^{-1}	85
	288 mA h g^{-1} after 120 cycles	2 A g^{-1}	
B-doped SnO_x/CNFs	670.2 mA h g^{-1} after 100 cycles	200 mA g^{-1}	86
	300 mA h g^{-1} after 80 cycles	2 A g^{-1}	
Cu-doped SnO_x/CNFs	743 mA h g^{-1} after 100 cycles	200 mA g^{-1}	87
	347 mA h g^{-1} after 1000 cycles	5 A g^{-1}	
SnO_2@PC/CTs	1045 mA h g^{-1} after 300 cycles	500 mA g^{-1}	88
	499 mA h g^{-1} after 1000 cycles	2A g^{-1}	
U-SnO_x/CNFs	608 mA h g^{-1} after 200 cycles	500 mA g^{-1}	80
	175 mA h g^{-1} after 40 cycles	5 A g^{-1}	

Table 1. Electrochemical performance of Sn-based composites with CNFs.

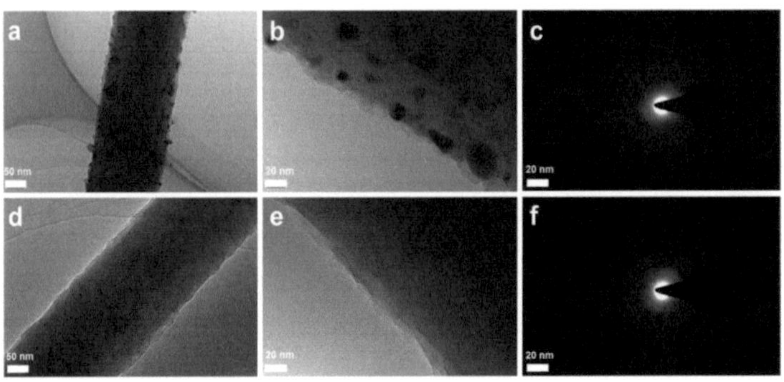

Figure 9. (a and b) HRTEM images of SnO_x/CNFs and (c) corresponding selected area electron diffraction (SAED) pattern; (d and e) HRTEM images of SnO_x–20%Cu/CNFs and (f) corresponding SAED pattern after 1000-cycle performance test at a current density of 2 A g^{-1} [86].

Highly enhanced performance of SnO_2 could be achieved by designing a novel 1D nanostructure. Liu et al [87] have designed and synthesized a novel fiber-in-tube hierarchical nanostructure of SnO_2@porous carbon in carbon tubes (denoted as SnO_2 @PC/CTs), with 1D

SnO_2@PC as the fibrous core and PDA-derived carbon tubes as the tubular shell, through N-doped carbon coating on electrospun hybrid nanofiber template and a post-etching technique (**Figure 10**). The internal PC skeleton could link and support SnO_2 nanoparticles for inhibiting the nanoparticle aggregation during cycling, while the external carbon protective shell could confine the volume expansion of SnO_2 for preserving the integrity of the overall electrode and facilitate electron and ion transport to the internal active materials. As a result, compared with SnO_2/CTs (without internal porous carbon skeleton), the SnO_2@PC/CT nanohybrids exhibit a higher reversible capacity of 1045 mA h g^{-1} at 0.5 A g^{-1} after 300 cycles and a high-rate cycling stability after 1000 cycles (Table 1) compared with those of SnO_2/CTs (without internal porous carbon skeleton). This unique 1D hierarchical nanostructure could be extended for improving other high-capacity metal oxides materials such as MnO/MnO_2, Fe_2O_3, and Co_3O_4.

3.3. Ti-based composite anodes with CNFs

TiO_2 has been regarded as a promising high-rate anode material due to its low cost, high working voltage, and structural stability during lithium insertion and extraction processes [88–93]. Bulk TiO_2 particle has poor ion and electron conductivity, which has limited its practical capacity and high-rate capability. So much attention has been paid to produce nanostructured

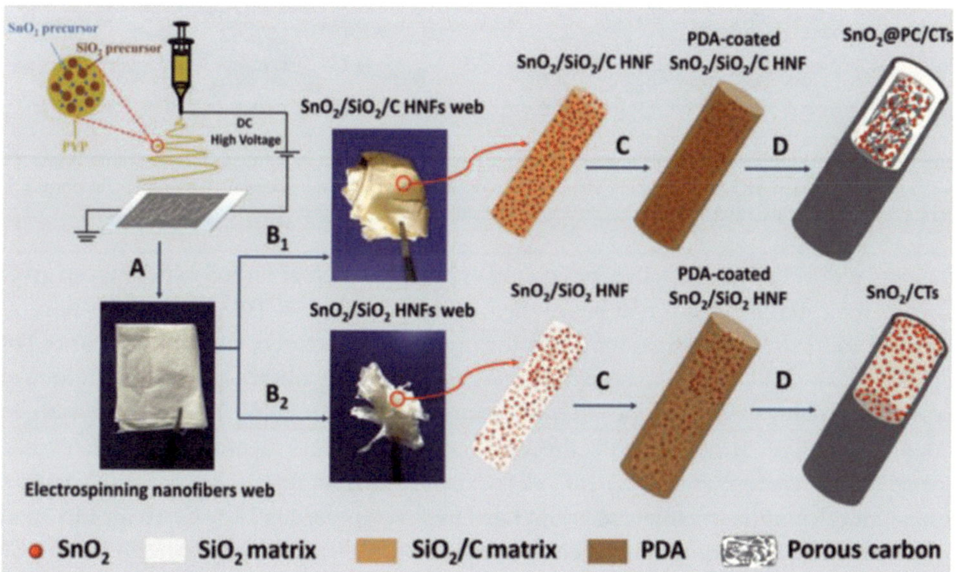

Figure 10. Schematic illustration on the preparation of SnO_2@PC/CT and SnO_2/CT nanohybrids: (A) preparing nanofiber web using electrospinning technique; (B) calcining electrospun nanofiber web (B_1) at 500°C for 1 h in air to obtain SnO_2/SiO_2/C HNF web and (B_2) at 600°C for 6 h in air to obtain SnO_2/SiO_2 HNF web (C) coating PDA on the surface of SnO_2/SiO_2/C HNFs or SnO_2/SiO_2 HNFs at room temperature; (D) carbonizing two types of PDA-coated HNFs at 600°C, and then etching SiO_2 to obtain SnO_2@PC/CTs with a fiber-in-tube hierarchical nanostructure or SnO_2/CTs with a particle-in-tube nanostructure [87].

and open-channeled TiO_2 materials, which can provide increased reaction active sites and short diffusion lengths for electron and lithium-ion transport [94–100].

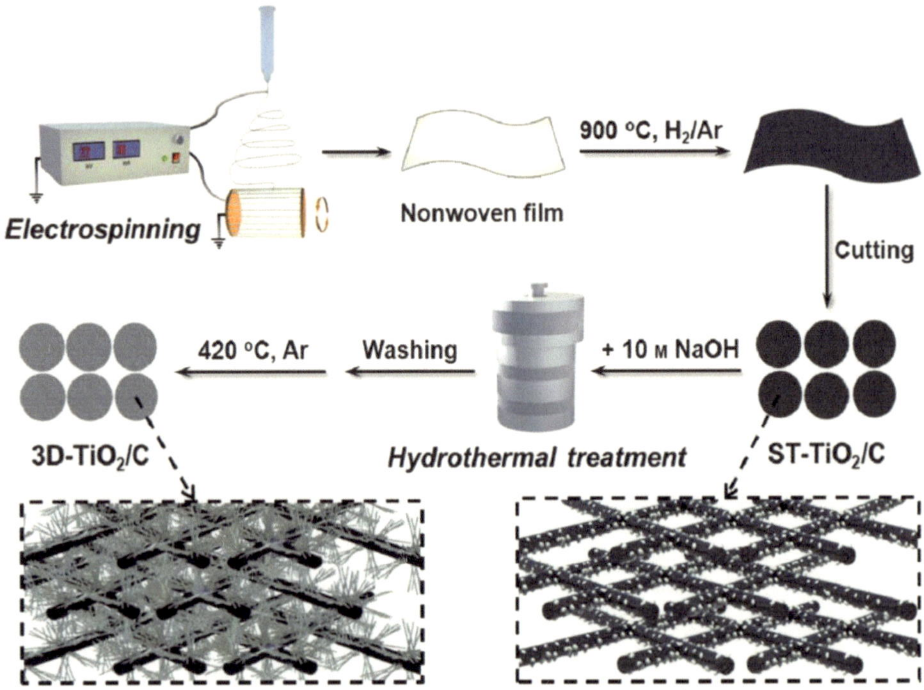

Figure 11. Schematic representation of the preparation for the 3D porous TiO_2 nanotube/carbon nanofiber architecture (ST–TiO_2/C: electrospun TiO_2/C nanofibers after calcination as a starting raw material; 3D-TiO_2/C: 3D porous TiO_2 nanotube/carbon nanofiber architecture prepared by a hydrothermal method) [106].

Moreover, tailoring these TiO_2 fires by coating or incorporation of carbon materials can greatly influence the capacity values and hence the battery performance [101, 102]. CNFs have also been used to load TiO_2 particles or fibers for improving electrochemical performance [103–106]. The TiO_2–CNF composite nanofibers were prepared by electrospinning technique and thermal treatment. Owing to the unique features of encapsulating TiO_2 nanocrystals into porous conductive carbon matrix, the composite nanofibers demonstrated an excellent electrochemical performance [103, 104]. A coaxial electrospinning technique combined with subsequent calcination treatment was also used to develop porous TiO_2–CNFs for LIB anodes [105]. In addition, a 3D porous architecture composed of TiO_2 nanotubes connected with a CNF matrix was successfully prepared by a hydrothermal method using electrospun rutile TiO_2/C nonwoven as the starting raw material (**Figure 11**) [106]. With its unique structure and connected conductive CNF network, the 3D architecture of the electrode resulted in superior rate performance: the reversible capacities were 214, 180, 138, and 112 mA h g^{-1} at the rate of 5, 10, 20, and 30 C, respectively. Additionally, the 3D structured electrode shows a very stable cycling performance, especially at a high rate of 30 C, without undergoing decay after 1000 cycles.

Spinel $Li_4Ti_5O_{12}$ has attracted particular attention for LIB application due to its nearly zero-strain characteristics [107] However, the practical application of $Li_4Ti_5O_{12}$ in LIBs is hampered by its poor natural electronic conductivity. In order to improve the conductivity of $Li_4Ti_5O_{12}$ materials, various approaches such as surface coating with conductive materials, e.g., Ag nanoparticles [108, 109], dispersion of $Li_4Ti_5O_{12}$ nanoparticles into a carbon matrix [110–112], and preparation of submicron or nanosized $Li_4Ti_5O_{12}$ [113, 114] aiming to significantly shorten the Li^+-diffusion length, etc., have been developed.

$Li_4Ti_5O_{12}$/carbon hybrid nanowebs consisting of interconnected nanofibers were prepared by a combination of electrospinning and subsequent thermal treatments [115, 116]. The as-prepared $Li_4Ti_5O_{12}$/carbon nanowebs exhibited high reversible charge stability and good cycling performance (166 mA h g^{-1} at 0.1 C). Highly porous $Li_4Ti_5O_{12}$/C nanofibers are successfully designed and prepared through electrospinning combined with a post-two-step annealing process [117]. The $Li_4Ti_5O_{12}$/C hybrid with a well-defined porous nanoarchitecture exhibits ultrahigh cycling rates and superior cycling stability. Mesoporous structures were also obtained by adding an amphiphilic triblock copolymer surfactant into a PVP solution, driving the self-assembly of a hydroxyl Li–Ti–O precursor to form mesopores after calcination [118]. Even better performances were obtained from combining dual-phase $Li_4Ti_5O_{12}$–TiO_2 with CNFs, prepared by immersing TiO_2/CNF in a LiOH solution at high temperatures. This imparted a pseudocapacitive effect, with a 204 mA h g^{-1} discharge capacity after 200 cycles at 100 mA g^{-1} from an initial capacity of ~220 mA h g^{-1} [119].

3.4. Other metal oxide/CNF composites

Many other metal oxide nanoparticles have also been incorporated into the CNF matrix for the usage of LIB anode materials. For example, the MnO_x particles, existed as MnO or Mn_3O_4, were incorporated into porous CNFs via electrospinning technique and subsequent heat treatment [120]. The porous MnO_x/C nanofibers experienced limited volume change with Li+ insertion/extraction because the ductile and strong C matrices suppressed the disintegration and aggregation of MnO_x. Compared with pure CNF anodes, the MnO_x/C exhibited larger charge and discharge capacities (542 mA h g^{-1} for MnO_x/C and 396 mA h g^{-1} for pure CNF at the 50th cycle) [121]. MnO_x was also incorporated into fibrous structures by the electrochemical deposition of MnO_x nanoparticles on PAN-based electrospun CNFs [122]. Similarly, Zhang et al. [123] prepared porous Co_3O_4–CNFs, which show an improved electrochemical performance compared to pure Co_3O_4 nanoparticles. C/Fe_3O_4 nanofibers with amorphous C structure and crystalline Fe_3O_4 particles were carbonized at a relatively low temperature (600°C), and showed high reversible capacity of 1007 mA h g^{-1} at the 80th cycle and excellent rate capability [124]. Recently, maghemite (γ-Fe_2O_3) nanoparticles were uniformly coated on CNFs by a hybrid synthesis procedure combining an electrospinning technique and hydrothermal method. Electrospun PAN nanofibers serve as a robust support for iron oxide precursors during the hydrothermal process and successfully limit the aggregation of nanoparticles at the following carbonization step (**Figure 12**) [125]. Such design not only increases the loading of Fe_2O_3 up to more than 60%, but also limits the aggregation of nanoparticles in the following carbonization step, which leads to a high reversible capacity of above 830 mA h g^{-1} after 40 cycles.

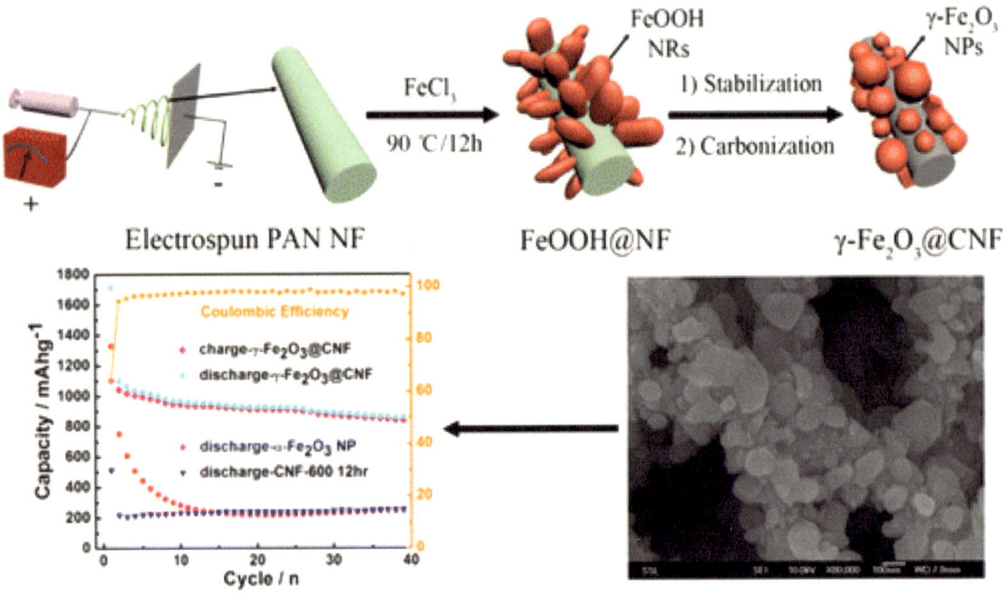

Figure 12. Schematic of the preparation of γ-Fe_2O_3@CNFs and its microstructure and performance [125].

4. Summary and perspective

In this chapter, the progress in electrospun CNFs and the composites with CNFs, which are used as LIB anode materials, has been summarized. Silicon, tin-based materials, and transition metal oxides are the candidates for the next generation anodes due to their expected high theoretical capacity, but suffer from some issues such as the vast volume change and low electronic conductivity, which could result in lower cycling stability and rate performance. Fortunately, these issues might be solved via composing with electrospun CNF matrix due to their superior mechanical properties and electrical conductivity as well as unique 1D nanostructure. More importantly, these CNF-based composite anodes with an interfibrous web structure could be directly used as anodes without any conductive agent and binder or current collectors, which can greatly reduce the inactive weight and cost of the cells, and significantly improve the electrochemical performance of LIBs [126, 127]. Further enhancement of electrochemical performance could be achieved by constructing controllable 1D nanostructures and doping various materials with CNF-based hybrid nanofibers.

In addition, sodium-ion batteries (SIBs) are new-emerged promising candidates for new battery systems especially for large-scale and long-term electric energy storage applications due to their cost advantages [5, 6, 128–133]. Currently, many researchers have paid more attention to electrospun materials for SIBs [134–142]. So far, research has mainly focused on the electrode materials based on electrospun 1D composite with CNFs. By constructing controllable 1D nanostructure to take full advantage of the electrospun materials including shorter diffusion pathways, high surface areas and porosities, good mechanical strengths, etc., one can find the wide use of the electrospun materials in commercial LIBs or SIBs, and even in flexible battery devices in the years ahead.

Author details

Yunhua Yu*, Yuan Liu and Xiaoping Yang

*Address all correspondence to: yuyh@mail.buct.edu.cn

State Key Laboratory of Organic–Inorganic Composites, Beijing University of Chemical Technology, Beijing, P. R. China

References

[1] D. Linden, T. B. Reddy. Handbook of Batteries, McGraw-Hill: Unite States, 2002.

[2] B. Scrosati, J. Garche. Lithium batteries: status, prospects and future. J. Power Sources 195 (2010) 2419.

[3] J. M. Tarascon, M. Armand. Issues and challenges facing rechargeable lithium batteries. Nature 414 (2001) 359.

[4] P. Poizot, S. Laruelle, S. Grugeon, L. Dupont, J. M. Tarascon. Nano-sized transition-metal oxides as negative-electrode materials for lithium-ion batteries. Nature 407 (2000) 496.

[5] R. Berthelot, D. Carlier, C. Delmas. Electrochemical investigation of the P2–NaxCoO2 phase diagram. Nat. Mater. 10 (2011) 74.

[6] N. Yabuuchi, M. Kajiyama, J. Iwatate, H. Nishikawa, S. Hitomi, R. Okuyama, R. Usui, Y. Yamada, S. Komaba. P2-type Nax[Fe1/2Mn1/2]O2 made from earth-abundant elements for rechargeable Na batteries. Nat. Mater. 11 (2012) 512.

[7] W.-J. Zhang. A review of the electrochemical performance of alloy anodes for lithium-ion batteries. J. Power Sources 196 (2011) 13.

[8] H. Wang, T. Abe, S. Maruyama, Y. Iriyama, Z. Ogumi, K. Yoshikawa. Graphitized Carbon Nanobeads with an Onion Texture as a Lithium-Ion Battery Negative Electrode for High-Rate Use. Adv. Mater. 17 (2005) 2857.

[9] F. Su, X. S. Zhao, Y. Wang, L. Wang, J. Y. Lee. Hollow carbon spheres with a controllable shell structure. J. Mater. Chem. 16 (2006) 4413.

[10] K. Zhang, X. Li, J. Liang, Y. Zhu, L. Hu, Q. Cheng, C. Guo, N. Lin, Y. Qian. Nitrogen-doped porous interconnected double-shelled hollow carbon spheres with high capacity for lithium ion batteries and sodium ion batteries. Electrochim. Acta 155 (2015) 174.

[11] X. X. Wang, J. N. Wang, H. Chang, Y. F. Zhang. Preparation of Short Carbon Nanotubes and Application as an Electrode Material in Li-Ion Batteries. Adv. Funct. Mater. 17 (2007) 3613.

[12] H. Gao, F. Hou, X. Zheng, J. Liu, A. Guo, D. Yang, Y. Gong. Electrochemical property studies of carbon nanotube films fabricated by CVD method as anode materials for lithium-ion battery applications. Vacuum 112 (2015) 1.

[13] S. Yoon, S. Lee, S. Kim, K.-W. Park, D. Cho, Y. Jeong. Carbon nanotube film anodes for flexible lithium ion batteries. J. Power Sources 279 (2015) 495.

[14] S.-H. Yoon, C.-W. Park, H. Yang, Y. Korai, I. Mochida, R. T. K. Baker, N. M. Rodriguez. Novel carbon nanofibers of high graphitization as anodic materials for lithium ion secondary batteries. Carbon 42 (2004) 21.

[15] X. Yan, D. Teng, X. Jia, Y. Yu, X. Yang. Improving the cyclability and rate capability of carbon nanofiber anodes through in-site generation of SiOx-rich overlayers. Electrochim. Acta 108 (2013) 196.

[16] H. Yue, F. Li, Z. Yang, J. Tang, X. Li, D. He. Nitrogen-doped carbon nanofibers as anode material for high-capacity and binder-free lithium ion battery. Mater. Lett. 120 (2014) 39.

[17] J. Tang, G. Chen, J. Yang, X. Zhou, L. Zhou, B. Huang. Silica-assistant synthesis of three-dimensional graphene architecture and its application as anode material for lithium ion batteries. Nano Energy 8 (2014) 62.

[18]] Y. Zhan, B. Zhang, L. Cao, X. Wu, Z. Lin, X. Yu, X. Zhang, D. Zeng, F. Xie, W. Zhang, J. Chen, H. Meng. Iodine doped graphene as anode material for lithium ion battery. Carbon 94 (2015) 1.

[19] D. Cai, S. Wang, L. Ding, P. Lian, S. Zhang, F. Peng, H. Wang. Superior cycle stability of graphene nanosheets prepared by freeze-drying process as anodes for lithium-ion batteries. J. Power Sources 254 (2014) 198.

[20] J. Zhang, Y.-S. Hu, J.-P. Tessonnier, G. Weinberg, J. Maier, R. Schlgl, D. S. Su. CNFs@CNTs: superior carbon for electrochemical energy storage. Adv. Mater. 20 (2008) 1450.

[21]] Z.-J. Fan, J. Yan, T. Wei, G.-Q. Ning, L.-J. Zhi, J.-C. Liu, D.-X. Cao, G.-L. Wang, F. Wei. Nanographene-constructed carbon nanofibers grown on graphene sheets by chemical

vapor deposition: high-performance anode materials for lithium ion batteries. ACS Nano 5 (2011) 2787.

[22] X. Zhang, S. Han, C. Fan, L. Li, W. Zhang. Hard carbon enveloped with graphene networks as lithium ion battery anode. Mater. Lett. 138 (2015) 259.

[23] C. Kim, K. S. Yang, M. Kojima, K. Yoshida, Y. J. Kim, Y. A. Kim, M. Endo. Fabrication of Electrospinning-Derived Carbon Nanofiber Webs for the Anode Material of Lithium-Ion Secondary Batteries. Adv. Funct. Mater. 16 (2006) 2393.

[24]] C. A. Bonino, L. Ji, Z. Lin, O. Toprakci, X. Zhang, S. A. Khan. Electrospun carbon-tin oxide composite nanofibers for use as lithium ion battery anodes. ACS Appl. Mater. Inter. 3 (2011) 2534.

[25] A. A. Madhavan, S. Kalluri, D. K. Chacko, T. A. Arun, S. Nagarajan, K. R. V. Subramanian, A. S. Nair, S. V. Nair, A. Balakrishnan. Electrical and optical properties of electrospun TiO2-graphene composite nanofibers and its application as DSSC photo-anodes. RSC Adv. 2 (2012) 13032.

[26] X. F. Wang, B. Ding, G. Sun, M. Wang, J. Y. Yu. Electro-spinning/netting: A strategy for the fabrication of three-dimensional polymer nano-fiber/nets. Prog. Mater. Sci. 58 (2013) 1173.

[27]] D. Li, Y. Xia. Fabrication of Titania Nanofibers by Electrospinning. Nano Lett. 3 (2003) 555.

[28] Z. M. Huang, Y. Z. Zhang, M. Kotaki, S. Ramakrishna. A review on polymer nanofibers by electrospinning and their applications in nanocomposites. Compos. Sci. Technol. 63 (2003) 2223.

[29]] A. Greiner, J. H. Wendorff. Inside Cover: A Molecular Solomon Link. Angew. Chem. Int. Ed. 46 (2007) 5670.

[30] Y. H. Wang, B. Li, Y. H. Liu, L. M. Zhang, Q. H. Zuo, L. F. Shi, Z. M. Su. Highly sensitive oxygen sensors based on Cu(I) complex-polystyrene composite nanofibrous membranes prepared by electrospinning. Chem. Commun. 39 (2009) 5868.

[31] H. G. Wang, Y. X. Li, L. Sun, Y. C. Li, W. Wang, S. Wang, S. F. Xu, Q. B. Yang. Electrospun novel bifunctional magnetic-photoluminescent nanofibers based on Fe2O3 nanoparticles and europium complex. J. Colloid Interface Sci. 350 (2010) 396.

[32] L. Xu, H. W. Song, B. Dong, Y. Wang, J. Chen, X. Bai. Preparation and Bifunctional Gas Sensing Properties of Porous In2O3-CeO2 Binary Oxide Nanotubes. Inorg. Chem. 49 (2010) 10590.

[33]] L. M. Li, X. M. Yin, S. Liu, Y. G. Wang, L. B. Chen, T. H. Wang. Electrospun porous SnO2 nanotubes as high capacity anode materials for lithium ion batteries. Electrochem. Commun. 12 (2010) 1383.

[34]] C. L. Zhang, S. H. Yu. Nanoparticles meet electrospinning: recent advances and future prospects. Chem. Soc. Rev. 43 (2014) 4423.

[35] H. L. Qu, S. Y. Wei, Z. H. Guo. Coaxial electrospun nanostructures and their applications. J. Mater. Chem. A 1 (2013) 11513.

[36] J. T. McCann, D. Li, Y. N. Xia. Electrospinning of nanofibers with core-sheath, hollow, or porous structures. J. Mater. Chem. 15 (2005) 735.

[37] J. C. Di, Y. Zhao, J. H. Yu. Fabrication of molecular sieve fibers by electrospinning. J. Mater. Chem. 21 (2011) 8511.

[38] S. Agarwal, A. Greiner, J. H. Wendorff. Electrospinning of Manmade and Biopolymer Nanofibers-Progress in Techniques, Materials, and Applications. Adv. Funct. Mater. 19 (2009) 1.

[39] L. F. Zhang, A. Aboagye, A. Kelkar, C. L. Lai, H. Fong. A review: carbon nanofibers from electrospun polyacrylonitrile and their applications. J. Mater. Sci. 49 (2014) 463.

[40] D. Li, Y. Xia. Electrospinning of Nanofibers: Reinventing the Wheel? Adv. Mater. 16 (2004) 1151.

[41] A. L. Yarin, E. Zussman, J. H. Wendorff, A. Greiner. Material encapsulation and transport in core-shell micro/nanofibers, polymer and carbon nanotubes and micro/nanochannels. J. Mater. Chem. 17 (2007) 2585.

[42] E. Teo, S. Ramakrishna. A review on electrospinning design and nanofibre assemblies. Nanotechnology 17 (2006) R89.

[43]] H. G. Wang, S. Yuan, D. -L. Ma, X.-B. Zhang, J.-M. Yan. Electrospun materials for lithium and sodium rechargeable batteries: from structure evolution to electrochemical performance. Energy Environ. Sci. 8 (2015) 1660.

[44] E. S. Pampal, E. Stojanovska, B. Simon, A. Kilic. A review of nanofibrous structures in lithium ion batteries. J. Power Sources 300 (2015) 199.

[45] M. Inagaki, Y. Yang, F. Y. Kang. Carbon Nanofibers Prepared via Electrospinning. Adv. Mater. 24 (2012) 2547

[46] Z. Dong, S. J. Kennedy, Y. Wu. Electrospinning materials for energy-related applications and devices. J. Power Sources 196 (2011) 4886.

[47]] N. A. Kaskhedikar, J. Maier. Lithium Storage in Carbon Nanostructures. Adv. Mater. 21 (2009) 2664.

[48] L. Dai, D. W. Chang, J.-B. Baek, W. Lu. Carbon Nanomaterials for Advanced Energy Conversion and Storage. Small 8 (2012) 1130.

[49]] X. D. Yan, D. H. Teng, X. L. Jia, Y. H. Yu, X. P. Yang. Improving the cyclability and rate capability of carbon nanofiber anodes through in-site generation of SiOx-rich overlayers. Electrochim. Acta 108 (2013) 196

[50] L. Feng, N. Xie, J. Zhong. Carbon Nanofibers and Their Composites: A Review of Synthesizing, Properties and Applications. Materials 7 (2014) 3919.

[51] J. Shin, W.-H. Ryu, K. -S. Park, Il-Doo Kim. Morphological Evolution of Carbon Nanofibers Encapsulating SnCo Alloys and Its Effect on Growth of the Solid Electrolyte Interphase Layer. ACS Nano 7 (2013) 7330.

[52] Y. Z. Wu, M. V. Reddy, B. V. R. Chowdari, S. Ramakrishna. Long-Term Cycling Studies on Electrospun Carbon Nanofibers as Anode Material for Lithium Ion Batteries. ACS Appl. Mater. Inter. 5 (2013) 12175.

[53] Y. H. Yu, Q. Yang, D. Teng, X. Yang, S. Ryu. Reticular Sn nanoparticle-dispersed PAN-based carbon nanofibers for anode material in rechargeable lithium-ion batteries. Electrochem. Commun. 12 (2010) 1187.

[54] L. Zou, L. Gan, F. Y. Kang, M. X. Wang, W. C. Shen, Z. H. Huang. Sn/C non-woven film prepared by electrospinning as anode materials for lithium ion batteries. J. Power Sources 195 (2010) 1216.

[55] P. Wang, D. Zhang, F. Ma, Y. Ou, Q. N. Chen, S. Xie, J. Li. Mesoporous carbon nanofibers with a high surface area electrospun from thermoplastic polyvinylpyrrolidone. Nanoscale 4 (2012) 7199.

[56]] X. Fan, L. Zou, Y.-P. Zheng, F.-Y. Kang, W.-C. Shen. Electrospinning Preparation of Nanosilicon/Disordered Carbon Composite as Anode Materials in Li-Ion Battery. Electrochem. Solid-State Lett. 12 (2009) A199.

[57] S.-X. Wang, L. Yang, L. P. Stubbs, X. Li, C. He. Lignin-Derived Fused Electrospun Carbon Fibrous Mats as High Performance Anode Materials for Lithium Ion Batteries. ACS Appl. Mater. Inter. 5 (2013) 12275.

[58] W. E. Tenhaeff, O. Rios, K. More, M. A. McGuire. Highly Robust Lithium Ion Battery Anodes from Lignin: An Abundant, Renewable, and Low-Cost Material. Adv. Funct. Mater. 24 (2014) 86.

[59] B. X. Liu, Y. H. Yu, J. Chang, X. J. Yang, D. Z. Wu, X. P. Yang. An enhanced stable-structure core-shell coaxial carbon nanofiber web as a direct anode material for lithium-based batteries. Electrochem. Commun. 13 (2011) 558.

[60] D. Nan, J.-G. Wang, Z.-H. Huang, L. Wang, W. Shen, F. Kang. Highly porous carbon nanofibers from electrospun polyimide/SiO2 hybrids as an improved anode for lithium-ion batteries. Electrochem. Commun. 34 (2013) 52.

[61]] B.-S. Lee, S.-B. Son, K.-M. Parka, W.-R. Yu, K.-H. Oh, S.-H. Lee. Anodic properties of hollow carbon nanofibers for Li-ion battery. J. Power Sources 199 (2012) 53.

[62] X. Li, X. B. Zhu, Y. Zhu, Z. Yuan, L. Si, Y. Qian. Porous nitrogen-doped carbon vegetable-sponges with enhanced lithium storage performance. Carbon 69 (2014) 515.

[63] P. Han, Y. Yue, L. Zhang, H. Xu, Z. Liu, K. Zhang, C. Zhang, S. Dong, W. Ma, G. Cui. Nitrogen-doping of chemically reduced mesocarbon microbead oxide for the improved performance of lithium ion batteries. Carbon 50 (2012) 1355.

[64] W. H. Shin, H. M. Jeong, B. G. Kim, J. K. Kang, J. W. Choi. Nitrogen-Doped Multiwall Carbon Nanotubes for Lithium Storage with Extremely High Capacity. Nano Lett. 12 (2012) 2283.

[65] Y. Liu, X. D. Yan, Y. H. Yu, X. P. Yang. Self-improving anodes for lithium-ion batteries: continuous interlamellar spacing expansion induced capacity increase in polydopaminederived nitrogen-doped carbon tubes during cycling. J. Mater. Chem. A. 3 (2015) 20880.

[66] L. W. Ji, X. W. Zhang. Manganese oxide nanoparticle-loaded porous carbon nanofibers as anode materials for high-performance lithium-ion batteries.Electrochem. Commun. 11 (2009) 795

[67] D. H. Teng, Y. H. Yu, P. W. Li, X. Bai. A topographically triplex-roughened Ti3O5/TiP2O7@MPCNFs hierarchical nanocomposite delivering synergistic lithium storage. RSC Adv. 3 (2013) 14237.

[68]] W. Li, M. Li, M. Wang, L. Zeng, Y. Yu. Electrospinning with partially carbonization in air: Highly porous carbon nanofibers optimized for high-performance flexible lithium-ion batteries. Nano Energy 13 (2015) 693.

[69]] L. Wang, C. X. Ding, L. C. Zhang, H. W. Xu, D. W. Zhang, T. Cheng, C. H. Chen. A novel carbon–silicon composite nanofiber prepared via electrospinning as anode material for high energy-density lithium ion batteries. J. Power Sources 195 (2010) 5052.

[70] H. S. Choi, J. G. Lee, H. Y. Lee, S. W. Kim, C. R. Park. Effects of surrounding confinements of Si nanoparticles on Si-based anode performance for lithium ion batteries. Electrochim. Acta 56 (2010) 790.

[71]] L. Ji, K.-H. Jung, A. J. Medford, X. Zhang. Electrospun polyacrylonitrile fibers with dispersed Si nanoparticles and their electrochemical behaviors after carbonization.J. Mater. Chem. 19 (2009) 4992.

[72] L. Ji, X. Zhang. Fabrication of porous carbon/Si composite nanofibers as high-capacity battery electrodes. Electrochem. Commun. 11 (2009) 1146.

[73] T. H. Wang, Y. M. Lee, B.-S. Kong, J.-S. Seo, J. W. Choi. Electrospun Core–Shell Fibers for Robust Silicon Nanoparticle-Based Lithium Ion Battery Anodes. Nano Lett. 12 (2012) 802.

[74] J. Kong, W. A. Yee, Y. Wei, L. Yang, J. M. Ang, S. L. Phua, S. Y. Wong, R. Zhou, Y. Dong, X. Li, X. Lu. Silicon nanoparticles encapsulated in hollow graphitized carbon nanofibers for lithium ion battery anodes. Nanoscale 5 (2013) 2967.

[75] Q. L. Wu, T. Tran, W. Q. Lu, J. Wu. Electrospun silicon/carbon/titanium oxide composite nanofibers for lithium ion batteries. J. Power Sources 258 (2014) 39.

[76] Y. Yu, L. Gu, C. B. Zhu, P. A. van Aken, J. Maier. Encapsulation of Sn@carbon Nanoparticles in Bamboo-like Hollow Carbon Nanofibers as an Anode Material in Lithium-Based Batteries. J. Am. Chem. Soc. 131 (2009) 15984.

[77] Y. Yu, L. Gu, C. L. Wang, A. Dhanabalan, P. A. van Aken, J. Maier. Encapsulation of Sn@carbon Nanoparticles in Bamboo-like Hollow Carbon Nanofibers as an Anode Material in Lithium-Based Batteries. Angew. Chem. Int. Ed. 48 (2009) 6485.

[78]] Z. X. Yang, G. D. Du, Z. P. Guo, X. B. Yu, S. Li, Z. X. Chen, P. Zhang, H. K. Liu. Plum-branch-like carbon nanofibers decorated with SnO2 nanocrystals. Nanoscale 2 (2010) 1011.

[79] X. S. Zhou, Z. H. Dai, S. H. Liu, J. C. Bao, Y.-G. Guo. Ultra-Uniform SnOx/Carbon Nanohybrids toward Advanced Lithium-Ion Battery Anodes. Adv. Mater. 26 (2014) 3943.

[80] B.-O. Jang, S.-H. Park, W.-J. Lee. J. Electrospun Co–Sn alloy/carbon nanofibers composite anode for lithium ion batteries. Alloys Compd. 574 (2013) 325.

[81] J.-C. Kim, D.-W. Kim. Electrospun Cu/Sn/C Nanocomposite Fiber Anodes with Superior Usable Lifetime for Lithium- and Sodium-Ion Batteries. Chem. Asian. J. 9 (2014) 3313.

[82] D. Kim, D. Lee, J. Kim, J. Moon. Electrospun Ni-Added SnO2–Carbon Nanofiber Composite Anode for High-Performance Lithium-Ion Batteries. ACS Appl. Mater. Inter. 4 (2012) 5408.

[83] Y. Liu, X. D. Yan, J. L. Lan, D. H. Teng, Y. H. Yu, X. P. Yang. Ti-doped SnOx encapsulated in Carbon nanofibers with enhanced lithium storage properties. Electrochim. Acta 137 (2014) 9.

[84] X. W. Liu, D. H. Teng, T. Li, Y. H. Yu, X. H. Shao, X. P. Yang. Phosphorus-doped tin oxides/carbon nanofibers webs as lithium-ion battery anodes with enhanced reversible capacity. J. Power Sources 272 (2014) 614.

[85] Q. Li, J. L. Lan, Y. Liu, Y. H. Yu, X. P. Yang. Carbon nanofiber-supported B2O3–SnOx glasses as anode materials for high-performance lithium-ion batteries. RSC Adv. 5 (2015) 89099.

[86] C. Chi, J. L. Lan, J. M. Sun, Y. Liu, Y. H. Yu, X. P. Yang. Amorphous Cu-added/SnOx/CNFs composite webs as anode materials with superior lithium-ion storage capability. RSC Adv. 5 (2015) 41210.

[87] Y. Liu, J. L. Lan, Q. Cai, Y. H. Yu, Y. H. Lin, X. P. Yang. Encapsulating Tin Dioxide@Porous Carbon in Carbon Tubes: A Fiber-in-Tube Hierarchical Nanostructure for Superior Capacity and Long-Life Lithium Storage. Part. Part. Syst. Charact. 32 (2015) 952.

[88]] S. Y. Huang, L. Kavan, I. Exnar, M. Grätzel. Rocking Chair Lithium Battery Based on Nanocrystalline TiO2 (Anatase). J. Electrochem. Soc. 142 (1995) L142.

[89] O. Wilhelm, S. E. Pratsinis, E. de chambrier, M. Crouzet, I. Exnar. Electrochemical performance of granulated titania nanoparticles. J. Power Sources 134 (2004) 197.

[90]] Z. L. Liu, L. Hong, B. Guo. Physicochemical and electrochemical characterization of anatase titanium dioxide nanoparticles. J. Power Sources 143 (2005) 231.

[91] M. Wagemaker, W. J. H. Borghols, F. M. Mulder. Large Impact of Particle Size on Insertion Reactions. A Case for Anatase LixTiO2. J. Am. Chem. Soc. 129 (2007) 4323.

[92] D. Cai, D. Li, S. Wang, X. Zhu, W. Yang, S. Zhang, H. Wang. High rate capability of TiO2/nitrogen-doped graphene nanocomposite as an anode material for lithium–ion batteries. J. Alloys Compd. 561 (2013), 54.

[93] C. Lai, G. R. Li, Y. Y. Dou, X. P. Gao. Mesoporous polyaniline or polypyrrole/anatase TiO2nanocomposite as anode materials for lithium-ion batteries. Electrochim. Acta. 55 (2010) 4567.

[94] S. J. Bao, Q. L. Bao, C. M. Li, Z. L. Dong. Novel porous anatase TiO2 nanorods and their high lithium electroactivity. Electrochem. Commun. 9 (2007) 1233.

[95] J. S. Chen, X. W. Lou. Anatase TiO2 nanosheet: An ideal host structure for fast and efficient lithium insertion/extraction. Electrochem. Commun. 11 (2009) 2332.

[96]] J. M. Szeifert, J. M. Feckl, D. F. Rohlfing, Y. J. Liu, V. Kalousek, J. Rathousky, T. Bein. Ultrasmall Titania Nanocrystals and Their Direct Assembly into Mesoporous Structures Showing Fast Lithium Insertion. J. Am. Chem. Soc. 132 (2010) 12605.

[97] J. H. Liu, J. S. Chen, X. F. Wei, X. W. Lou, X. W. Liu. Sandwich-Like, Stacked Ultrathin Titanate Nanosheets for Ultrafast Lithium Storage. Adv. Mater. 23 (2011) 998.

[98] J. Wang, Y. K. Zhou, Y. Y. Hu, R. O. Hayre, Z. P. Shao. Facile Synthesis of Nanocrystalline TiO2 Mesoporous Microspheres for Lithium-Ion Batteries. J. Phys. Chem. C. 115 (2011) 2529.

[99] M. V. Reddy, R. Jose, T. H. Teng, B. V. R. Chowdari, S. Ramakrishna. Preparation and electrochemical studies of electrospun TiO2nanofibers and molten salt method nanoparticles. Electrochim. Acta 55 (2010) 3109.

[100]] H. W. Lu, W. Zeng, Y. S. Li, Z. W. Fu. Fabrication and electrochemical properties of three-dimensional net architectures of anatase TiO2 and spinel Li4Ti5O12 nanofibers. J. Power Sources 164 (2007) 874.

[101] P. Zhu, Y. Wu, M. V. Reddy, A. S. Nair, B. V. R. Chowdari, S. Ramakrishna. Long term cycling studies of electrospun TiO 2 nanostructures and their composites with MWCNTs for rechargeable Li-ion batteries. RSC Adv. 2 (2012) 531.

[102] M. H. Ryu, K. N. Jung, K. H. Shin, K. S. Han, S. Yoon. High performance N-doped mesoporous carbon decorated TiO2 nanofibers as anode materials for lithium-ion batteries. J. Phys. Chem. C. 17 (2013) 8092.

[103] X. J. Yang, D. H. Teng, B. X. Liu, Y. H. Yu, X. P. Yang. Nanosized anatase titanium dioxide loaded porous carbon nanofiber webs as anode materials for lithium-ion batteries. Electrochem. Commun. 13 (2011) 1098.

[104] Z. X. Yang, G. D. Du, Q. Meng, Z. P. Guo, X. B. Yu, Z. X. Chen, T. L. Guo, R. Zeng. Synthesis of uniform TiO2@carbon composite nanofibers as anode for lithium ion batteries with enhanced electrochemical performance. J. Mater. Chem. (2012) 5848.

[105] X. Y. Li, Y. M. Chen, L. M. Zhou, Y.-W. Mai, H. T. Huang. Exceptional electrochemical performance of porous TiO 2–carbon nanofibers for lithium ion battery anodes. J. Mater. Chem. A. 2 (2014) 3875.

[106]] B. Zhao, S. Jiang, C. Su, R. Cai, R. Ran, M. O. Tad′eb, Z. P. Shao. A 3D porous architecture composed of TiO 2 nanotubes connected with a carbon nanofiber matrix for fast energy storage. J. Mater. Chem. A. 1 (2013) 12310.

[107] N. Li, H. Ge, D. Y. Li, C. S. Dai, D. L. Wang. Electrochemical characteristics of spinel Li4Ti5O12 discharged to 0.01 V. Electrochem. Commun. 10 (2008) 719.

[108] Z. Y. Wen, S. H. Huang, X. J. Zhu, Z. H. Gu. Preparation and electrochemical performance of Ag doped Li4Ti5O12. Electrochem. Commun. 6 (2004) 1093.

[109]] Z. Y. Wen, S. H. Huang, J. C. Zhang, Z. H. Gu, X. H. Xu. Li4Ti5O12/Ag composite as electrode materials for lithium-ion battery. Solid State Ionics 177 (2006) 851.

[110] J. Huang, Z. Jiang. The preparation and characterization of Li4Ti5O12/carbon nanotubes for lithium ion battery. Electrochim. Acta 53 (2008) 7756.

[111] Y. P. Wu, G. J. Wang, J. Gao, L. J. Fu, N. H. Zhao, T. Takamura. Preparation and characteristic of carbon-coated Li4Ti5O12 anode material. J. Power Sources. 174 (2007) 1109.

[112] L. Cheng, X. L. Li, H. J. Liu, H. M. Xiong, P. W. Zhang, Y. Y. Xia. Carbon-coated Li4Ti5O12 as a high rate electrode material for Li-ion intercalation. J. Electrochem. Soc. 154 (2007) A692.

[113] J. Kim, D. H. Kim, Y. S. Ahn. Polyol-mediated synthesis of Li4Ti5O12 nanoparticle and its electrochemical properties. Electrochem. Commun. 7 (2005) 1340.

[114] J. Kim, J. Cho. Spinel Li4Ti5O12 nanowires for high-rate Li-ion intercalation electrode. Electrochem. Solid Struct. 10 (2007) A81.

[115] Z. Yang, Q. Meng, Z. Guo, X. Yu, T. Guo, R. Zeng. Highly reversible lithium storage in uniform Li4Ti5O12/carbon hybrid nanowebs as anode material for lithium-ion batteries. Energy 55 (2013) 925.

[116] B. Guo, Y. Li, Y. Yao, Z. Lin, L. Ji, G. Xu, Yi. Liang, Q. Shi, X. Zhang. Electrospun Li4Ti5O12/C composites for lithium-ion batteries with high rate performance. Solid State Ionics 204–205 (2011) 61.

[117] H. Xu, X. Hu, Y. Sun, W. Luo, C. Chen, Y. Liu, Y. Huang. Highly porous Li4Ti5O12/C nanofibers for ultrafast electrochemical energy storage. Nano Energy 10 (2014) 163.

[118] J. Wang, L. Shen, H. Li, B. Ding, P. Nie, H. Dou, X. Zhang. A facile one-pot synthesis of TiO2/nitrogen-doped reduced graphene oxide nanocomposite as anode materials for high-rate lithium-ion batteries. J. Alloys Compd. 587 (2014) 171.

[119]] J. Sun, D. Teng, Y. Liu, C. Chi, Y. Yu, J.-L. Lan, X Yang. Enhanced lithium storage capability of a dual-phase Li4Ti5O12–TiO2–carbon nanofiber anode with interfacial pseudocapacitive effect. RSC Adv. 4 (2014) 48632.

[120]] L. Ji, A. J. Medford, X. Zhang. Porous carbon nanofibers loaded with manganese oxide particles: Formation mechanism and electrochemical performance as energy-storage materials. J. Mater. Chem. 19 (2009) 5593.

[121] L. Ji, X. Zhang. Manganese oxide nanoparticle-loaded porous carbon nanofibers as anode materials for high-performance lithium-ion batteries. Electrochem. Commun. 11 (2009) 795.

[122] Z. Lin, L. Ji, M. D. Woodroof, X. Zhang. Electrodeposited MnOx/carbon nanofiber composites for use as anode materials in rechargeable lithium-ion batteries. J. Power Sources 195 (2010) 5025.

[123] P. Zhang, Z. P. Guo, Y. D. Huang, D. Z. Jia, H. K. Liu. Synthesis of Co3O4/carbon composite nanowires and their electrochemical properties. J. Power Sources 196 (2011) 6987.

[124] L. Wang, Y. Yu, P. C. Chen, D. W. Zhang, C. H. Chen. Electrospinning synthesis of C/ Fe3O4 composite nanofibers and their application for high performance lithium-ion batteries. J. Power Sources 183 (2008) 717.

[125] Y. Z. Wu, P. N. Zhu, M. V. Reddy, B. V. R. Chowdari, S. Ramakrishna. Maghemite nanoparticles on electrospun CNFs template as prospective lithium-ion battery anode. ACS Appl. Mater. Inter. 6 (2014) 1951.

[126] G. M. Zhou, F. Li, H.-M. Cheng. Progress in flexible lithium batteries and future prospects.Energy Environ. Sci. 7 (2014) 1307.

[127] L. Li, Z. Wu, S. Yuan, X.-B. Zhang. Advances and challenges for flexible energy storage and conversion devices and systems. Energy Environ. Sci. 7 (2014) 2101.

[128] Y. H. Liu, Y. H. Xu, Y. J. Zhu, J. N. Culver, C. A. Lundgren, K. Xu, C. S. Wang. Tin-Coated Viral Nanoforests as Sodium-Ion Battery Anodes. ACS Nano 7 (2013) 3627.

[129] Y. H. Xu, Y. J. Zhu, Y. H. Liu, C. S. Wang. Electrochemical performance of porous carbon/tin composite anodes for sodium-ion and lithium-ion batteries. Adv. Energy Mater. 3 (2013) 128.

[130] Z. L. Jian, C. C. Yuan, W. Z. Han, X. Lu, L. Gu, X. K. Xi, Y.-S. Hu, H. Li, W. Chen, D. F. Chen, Y. Ikuhara, L. Q. Atomic Structure and Kinetics of NASICON NaxV2(PO4)3 Cathode for Sodium-Ion Batteries.Chen. Adv. Funct. Mater. 24 (2014) 4265.

[131] S. W. Wang, L. J. Wang, Z. Q. Zhu, Z. Hu, Q. Zhao. All Organic Sodium-Ion Batteries with Na4C8H2O6†.J. Chen. Angew. Chem., Int. Ed. 53 (2014) 5892.

[132] J. F. Qian, Y. Xiong, Y. L. Cao, X. P. Ai, H. X. Yang. Synergistic Na-storage reactions in Sn4P3 as a high-capacity, cycle-stable anode of Na-ion batteries.Nano Lett. 14 (2014) 1865.

[133] D. Y. W. Yu, P. V. Prikhodchenko, C. W. Mason, S. K. Batabyal, J. Gun, S. Sladkevich, A. G. Medvedev, O. Lev. High-capacity antimony sulphide nanoparticle-decorated graphene composite as anode for sodium-ion batteries. Nat. Commun. 4 (2013) 2922.

[134] T. Q. Chen, Y. Liu, L. K. Pan, T. Lu, Y. F. Yao, Z. Sun, D. H. C. Chua, Q. Chen. Electrospun carbon nanofibers as anode materials for sodium ion batteries with excellent cycle performance. J. Mater. Chem. A. 2 (2014) 4117.

[135] L. Wu, X. H. Hu, J. F. Qian, F. Pei, F. Y. Wu, R. J. Mao, X. P. Ai, H. X. Yang, Y. L. Cao. Sb–C nanofibers with long cycle life as an anode material for high-performance sodium-ion batteries. Energy Environ. Sci. 7 (2014) 323.

[136] L. W. Ji, M. Gu, Y. Y. Shao, X. L. Li, M. H. Engelhard, B. W. Arey, W. Wang, Z. M. Nie, J. Xiao, C. M. Wang, J.-G. Zhang, J. Liu. Controlling SEI Formation on SnSb-Porous Carbon Nanofibers for Improved Na Ion Storage. Adv. Mater. 26 (2014) 2901.

[137] C. B. Zhu, X. K. Mu, P. A. van Aken, Y. Yu, J. Maier. Single-Layered Ultrasmall Nanoplates of MoS2 Embedded in Carbon Nanofibers with Excellent Electrochemical Performance for Lithium and Sodium Storage. Angew. Chem., Int. Ed. 53 (2014) 2152.

[138] J. Liu, K. Tang, K. P. Song, P. A. van Aken, Y. Yu, J. Maier. Tiny Li4Ti5O12 nanoparticles embedded in carbon nanofibers as high-capacity and long-life anode materials for both Li-ion and Na-ion batteries. Phys. Chem. Chem. Phys. 15 (2013) 20813.

[139] J. Liu, K. Tang, K. P. Song, P. A. van Aken, Y. Yu, Electrospun Na3V2(PO4)3/C nanofibers as stable cathode materials for sodium-ion batteries. J. Maier. Nanoscale. 6 (2014) 5081.

[140] T. H. Hwang, D. S. Jung, J.-S. Kim, B. G. Kim, J. W. Choi. One-Dimensional Carbon–Sulfur Composite Fibers for Na–S Rechargeable Batteries Operating at Room Temperature. Nano Lett. 13 (2013) 4532.

[141] W. H. Li, L. C. Zeng, Z. Z. Yang, L. Gu, J. Q. Wang, X. W. Liu, J. X. Cheng, Y. Yu. Free-standing and binder-free sodium-ion electrodes with ultralong cycle life and high rate performance based on porous carbon nanofibers. Nanoscale 6 (2014) 693.

[142] Y. J. Zhu, X. G. Han, Y. H. Xu, Y. H. Liu, S. Y. Zheng, K. Xu, L. B. Hu, C. S. Wang. Electrospun Sb/C Fibers for a Stable and Fast Sodium-Ion Battery Anode. ACS Nano 7 (2013) 6378.

Chapter 2

Capacity Optimization Nanotechnologies for Enhanced Energy Storage Systems

Natasha Ross and Emmanuel I. Iwuoha

Additional information is available at the end of the chapter

http://dx.doi.org/10.5772/62319

Abstract

Rechargeable lithium-ion battery (LIB) cathodes consist of transition metal oxide material, which reversibly (de)intercalates lithium at a high potential difference versus a carbon anode. Manganese oxide cathode material offers lower cost and toxicity than the normally used cobalt. However, $LiMn_2O_4$ suffers from capacity fading, Mn dissolution at high temperatures, and poor high rate capability. Its ultimate performance, however, depends on the morphology and electrochemical properties. In this work, Au alloyed with Fe, Pd, and Pt, respectively, was synthesized and used to improve the microstructure and catalytic activities by functionalizing $LiMn_2O_4$ via a coprecipitation calcination method. The pristine $LiMn_2O_4$ and modified materials were examined using a combination of spectroscopic and microscopic techniques along with in-detail galvanostatic charge–discharge tests. Microscopic results revealed that the modified composite cathode materials had high phase purity, highly crystallized particles, and more regular morphological structures with narrow size distributions. Galvanostatic charge–discharge testing indicated that the initial discharge capacities of $LiM_xMn_{2-x}O_4$ at 0.1 C for $M_{0.02}$=PtAu, FeAu, and PdAu were 147, 155.5, and 160.2 mAh g^{-1}, respectively. The enhancement of the capacity retention and higher electrode coulombic efficiency of the modified materials were significant, especially at high C rate. At enlarged cycling potential ranges, the $Li(M)_{0.02}Mn_{1.98}O_4$ samples delivered relevant discharge capacities (70, 80, and 90 mAh g^{-1}) compared to $LiMn_2O_4$ (45 mAh g^{-1}).

Keywords: Energy storage, Capacity, Nanotechnology, Cathode, Electroanalytical chemistry

1. Introduction

Among various energy storage technologies, rechargeable lithium-ion batteries (LIBs) are considered as an effective and reliable solution to the increasing demand for high-energy density power sources. The high specific energy and power available from LIBs and the prospect to charge and discharge them hundreds of times are the reason for their key importance in electronic portable devices and future development of hybrid vehicles.

When a battery is charged or discharged, the redox reactions change the molecular or crystalline structure of the electrode materials, which often affects their stability, and hence require frequent replacement after several charge-discharge cycles. LIBs weigh less, take less space for a given energy delivery, and offer higher-energy densities than established Ni-Cd and Ni-MH batteries [1]. However, to acquire an established role in the commercial sector, LIBs require improved power density (W/kg), which depends primarily on the performance of the active materials. Li^+ cells operate by a mechanism whereby Li^+ shuttle between two host electrodes, commonly referred to as "insertion electrodes". An insertion compound has two specific properties: the guest ions, Li^+, are mobile between sites in the host network and can be removed from or added to the transition metal oxides host network, thus varying the guest ion concentration [2]. This process is known as intercalation and deintercalation of electrons, occurring at the insertion compound. For transition metal oxides or other chalcogenides, Li^+ occupy sites surrounded by oxygen or chalcogen ions. The sites available are determined by the host structure.

The theoretical capacity, denoted *Q*, of the host material can be calculated from Faraday's first law of electrochemistry, which states that 1 gram equivalent weight of a material will deliver 96,487 C (or 26.8 Ah). For $LiMn_2O_4$, the equivalent weight (*M*) is 180.8 g/mol, having a theoretical capacity of 26.8/180.8=148 mAh g^{-1}.

In LIB, the anode is well optimized and little improvements can be gained in terms of design changes. The cathode, however, shows promise for further enhancements, i.e., improved capacity and stability. Therefore, most research studies are focused on further development of the cathode active materials.

Cathode materials are typically oxides of transition metals, which can undergo oxidation to higher valences when lithium is removed. Among numerous transition metal oxides, $LiMn_2O_4$ has emerged as one of the promising candidates because of its 3D Li^+ diffusion, low cost, abundance, nontoxicity, $Mn^{3+/4+}$ redox couple, and excellent rate capability. $LiMn_2O_4$, however, suffers from low conductivity and capacity fading during cycling. $LiMn_2O_4$ adopts a (3D) structure described as a cubic close packing (ccp) of oxygen atoms with Mn occupying half of the octahedral and Li occupying an eighth of the tetrahedral sites referring to the 16d and 8a sites ($[Li]_{tet}[Mn_2]_{Oct}O_4$), respectively (**Figure 1a**). The $LiMn_2O_4$ morphology corresponds to that of a single crystal with a cubic structure (**Figure 1b**). The scanning confocal electron microscopy (SCEM) image (**Figure 1c**) exhibiting a negative feedback mode verifies the insulating nature of $LiMn_2O_4$. This alternately hinders the diffusion of the mediator to the electrode [3] and contributes to the poor electrochemical performance of the cathode.

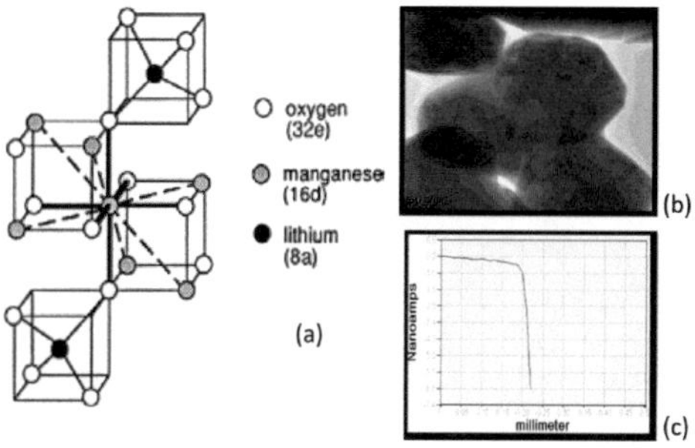

Figure 1. Spinel structure showing the MnO_6 octahedra and Li 8a tetrahedral positions of $LiMn_2O_4$ (a) [4] with its corresponding transmission electron microscopy (TEM; b) and SCEM (c) graphs.

$LiMn_2O_4$ with its poor stability and high rate performance cannot satisfy high-power applications. The reason for this may due to HF generated during cycling when using $LiPF_6$-based electrolyte, which is responsible for the dissolution of manganese. In addition, Mn^{3+} is considered to be the main source of the dissolution of manganese via Hunter's reaction [5]:

$$4H^{+} + 2LiMn^{3+}Mn^{4+}O_4 \rightarrow 3(\lambda - MnO_2) + Mn^{2+} + 2Li^{+} + 2H_2O$$

The protonated λ-MnO_2 cannot act as a host for Li^+ during cycling because of the strong binding energy of the protons to the oxygen sites around the 16d Mn [6] . This reduces conductance and blocks the transport of electrons or ions, reducing the electrochemical performance of $LiMn_2O_4$.

To solve partially or completely the capacity fading problems, coating the surface of $LiMn_2O_4$ is the easiest functionalization route and offers a suitable approach to avoid the dissolution of Mn^{2+} [7]. Many research studies have confirmed the importance of the structural surface features of cathode materials for their electrochemical performance. It is claimed that, when the surface of cathode materials is coated with oxides and other catalytically active materials, the coatings prevent direct contact with the electrolyte solution, suppress phase transition, improve the structural stability, and decrease the disorder of cations in crystal sites. As a result, side reactions and heat generation during cycling are minimal.

The aim of this study is synthesis and characterization of novel transition metal alloy surface-modified spinel $LiMn_2O_4$ nanocathode materials with high rate discharge capabilities. The coating material consists of Au alloyed with Fe, Pt, and Pd nanoparticles, respectively. The components were selected based on their ability to provide high electronic conductivity where needed and their potential to act as a chemical catalyst for electrochemical reactions at the nanoscale [8]. These alloys are relatively less studied when compared to other transition metal alloys commonly used to form network structures with $LiMn_2O_4$. The modified cathode allows the consumption of smaller amounts of expensive metals, boosting both the catalytic and economic appeals. Moreover, the synergy within the novel composite cathode materials can provide a better connecting network for electron diffusion due to a shortened transportation path [9] and enhanced phase transition kinetics of Li^+ ion intercalation/deintercalation, which may be ascribed to the nanoscale structure [10]. Emerging applications have therefore steered Li^+ electrode materials in the direction of nanomaterials [11]. The effect of surface functionalization on the conductivity, stability, and high rate performances [12] of $LiMn_2O_4$ was probed by spectroscopic and microscopic techniques along with galvanostatic charge–discharge analyses.

1.1. Significance of nanocatalysis

Nanocatalysis is when nanoparticles are used to catalyze chemical reactions and allow for rapid and selective chemical transformations. Nanomaterials can increase the efficiency of the energy storage and conversion processes as well as device design and performance. The application of bimetallic nanoparticles as catalysts is one of the most active areas of nanoscience [13]. Bimetallic nanoparticles are the combination of two metals in the nanometer range. They can be classified into four types of mixing patterns: core-shell nanoparticles, subcluster nanoparticles, mixed nanoparticles, and multishell nanoparticles [14]. The combination of the two different metals gives rise to an enhancement of specific properties. These properties may be different to those of pure elemental particles and can include unique size-dependent optical, electronic, and catalytic effects [15]. Bimetallic nanocatalysts provide a way to use smaller amounts of an expensive catalyst material, by using a less expensive metal for the core material. For transition metal nanoparticles, the decrease in size to the nanometer length scale increases the surface-to-volume ratio. This, together with the ability to make them in different sizes and shapes, makes them potentially useful in the field of catalysis [16]. It has been proven that the most favorable size for platinum-based electrocatalysis is in the range of 2 to 4 nm [17]. Bimetallic nanoparticles have been proven successful in enhancing the selectivity and rates of many targeted reactions. For example, the hydrogenation rate of simple olefins is increased in the presence of palladium catalysts containing 20% gold when compared to monometallic palladium catalysis [18]. In this study, we investigate how chemically depositing Au-Fe, Au-Pd, and Au-Pt can enhance the conductivity and cyclability of LiMn2O4.

2. Experimental method

Pristine spinel-$LiMn_2O_4$ cathode was prepared following a coprecipitation method [19]. Stoichiometric amounts of LiOH and ($MnAc_2.4H_2O$) with a cationic ratio of Li/Mn=1:2 were dissolved in deionized water and thoroughly mixed. This was followed by evaporation at 100ºC for 10 h. The precursor was further preheated at 400ºC for 1 h and then calcined at 800ºC for 20 h in air to form $LiMn_2O_4$. Subsequently, a multiple-phase emulsion-assisted approach was used to prepare Au-Fe, Au-Pd, and Au-Pt bimetallic nanoparticles with narrow size distributions. This process uses metallic acetates as the metal precursors. For functionalization of $LiMn_2O_4$, the nanoalloy was added to deionized water and allowed to dissolve completely. The crystalline $LiMn_2O_4$ was added to the alloy solution and heated until the solvent evaporates. This was then followed by calcination at 550ºC for 10 h in air. Hereby, the impurities were removed and Mn^{3+} concentration prone to disproportion was reduced and a composite of $Li(M)_xM_{2-x}O_4$ (x=0.02) was constructed.

2.1. Battery assembly

The tested cell consisted of the cathode and a lithium metal anode separated by Celgard microporous membranes to avoid internal short circuit. The electrolyte used was 1 M $LiPF_6$ ethylene carbonate (EC)-dimethyl carbonate (DMC; 1:2, by volume; Ube Chemicals, Japan) solution. All the assembling of the cell was carried out in an argon-filled dry box.

3. Results and discussion

The micrographs of pure and modified $LiMn_2O_4$ samples are shown in **Figure 2**. The pure $LiMn_2O_4$ (a) appear to have primary particles around 50 nm and have the expected spinel shape. The secondary particles of $LiMn_2O_4$ are about 100 nm, which are glomeration congregated tightly by primary particles, indicating that the crystals of the spinel $LiMn_2O_4$ grow well and have interparticle boundaries. The micrograph of $LiFeAu_{0.02}Mn_{1.98}O_4$ (b) shows two types of particles with different particle sizes. The larger particles are that of the pristine sample, exemplified by an octahedral shape. The FeAu nanoparticle surface feature corresponds to cubic- or hexagonal-shaped particles of about 20 nm, which are heterogeneously dispersed across the pristine $LiMn_2O_4$ particles. $LiPdAu_{0.02}Mn_{1.98}O_4$ (c) shows well-dispersed PdAu nanoparticles. The $LiPdAu_{0.02}Mn_{1.98}O_4$ nanoparticles retained a well-developed octahedral structure with sharp edges after surface treatment and exhibited particle sizes in the range of 20–50 nm. The micrograph of $LiPtAu_{0.02}Mn_{1.98}O_4$ (d) shows smoother particles. The surface of $LiPtAu_{0.02}Mn_{1.98}O_4$ appears infused with the PtAu nanoparticles having a bumpy morphology with only some aggregates [20]. All the coated $LiMn_2O_4$ samples have increased surface area, which favors the penetration of electrolyte, decreasing the diffusion length of Li^+ and electrons, and improves the overall activity of the cathode.

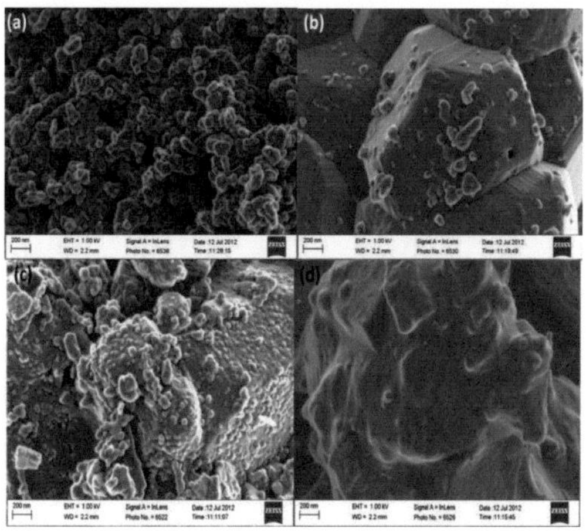

Figure 2. SEM micrographs of $LiMn_2O_4$ (a), $LiFeAu_{0.02}Mn_{1.98}O_4$ (b), $LiPdAu_{0.02}Mn_{1.98}O_4$ (c), and $LiPtAu_{0.02}Mn_{1.98}O_4$ (d) calcined at 880°C.

Atomic force microscopy (AFM) was used to characterize samples at the microscope level, with accurate resolution ranging from 100 to <1 μm. **Figure 3** shows the AFM surface morphology images of $LiMn_2O_4$ and $Li(M)_{0.02}Mn_{1.98}O_4$ particles. Compared to scanning electron microscopy (SEM), AFM provides extraordinary topographic contrast direct height measurements and unobstructed views of surface features. The darker regions are indicative of the partial heterogeneous surface coverage. The $LiMn_2O_4$ (a) image reveals clear globular features. It is evident that this surface yields a large degree of surface roughness. The size distribution was broad and the maximum shifted towards an average diameter of 60 nm. It has been known that the particle size influences directly on battery performance; hence, materials generally employed for battery application are >1 μm. To apprehend the high rate capabilities of the cathode, it is necessary to employ active materials with particles in the nanometer range.

The $LiFeAu_{0.02}Mn_{1.98}O_4$ (b) sample yields uniform surface roughness. This feature may be attributed to the FeAu coating layer, which attributed to compact surface architecture. Evidently, the particle size distribution has narrowed down and the maximum has shifted towards an average diameter of 40 nm. Based on prior research [21], it is clear that the smaller the crystallite size, the bigger the specific surface area, hence permitting favourable conditions for improved cycling performance. $LiPdAu_{0.02}Mn_{1.98}O_4$ (c) revealed uniform surface roughness, and the particle sizes distribution is narrow with an average maximum diameter of 30 nm. A new phase possessed by $LiPtAu_{0.02}Mn_{1.98}O_4$ is evident in (d). The observed changes could be the result of the film volume increase, which is due to the uptake of PtAu onto the $LiMn_2O_4$ matrix [22]. AFM characterization confirmed that the PtAu alloy is well adherent to the surface. This agrees with the SEM image showing PtAu particle infusion. The surface feature may be beneficial for more rapid Li^+ transportation facilitating better high rate performances, including low polarization and better structural stability.

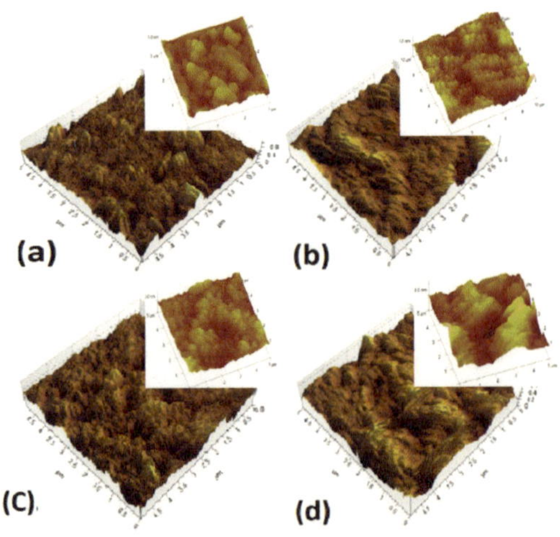

Figure 3. AFM images of $LiMn_2O_4$ (a), $LiFeAu_{0.02}Mn_{1.98}O_4$ (b), $LiPdAu_{0.02}Mn_{1.98}O_4$ (c), and $LiPtAu_{0.02}Mn_{1.98}O_4$ (d).

Figure 4 shows the X-ray diffraction (XRD) peaks of $LiMn_2O_4$ and $Li(M)_{0.02}Mn_{1.98}O_4$ cathode materials. All the peaks correspond to a single phase of cubic spinel structure with a space group *Fd3m* in which the Li^+ occupy the tetrahedral (8a) site and manganese and the substituted metal occupy the octahedral (16d) site. All the peaks were indexed as the spinel phase (JCPDS: 35-0782). These results indicate that $LiMn_2O_4$ retains its spinel structure after modification and the addition of the bimetalic alloys in the preparation process does not alter the phase purity of the product. The main diffraction peaks of cubic spinel $LiMn_2O_4$ phase, such as (111), (311), and (400), is well developed. No additional peak was observed for $Li(M)_{0.02}Mn_{1.98}O_4$; however, the lattice constants change slightly compared to pristine spinel powder from 8.2609 to 8.2600 Å. The ionic radius of Mn^{4+} is smaller than that of Mn^{3+}. Therefore, the decrease of lattice constant is indicative of an increase in Mn^{4+} ion concentration in the spinel [23]. The smaller MnO_6 octahedra suggest a stable MnO_6 framework and lower activation energy, which supports high rate electrochemical cycling [24]. Concomitantly, the coated samples with almost the same structure as $LiMn_2O_4$ support easy Li^+ diffusion during the charge-discharge process. This finding was validated by both solid-state nuclear magnetic resonance (NMR) magic angle spinning (MAS) and cycliability studies.

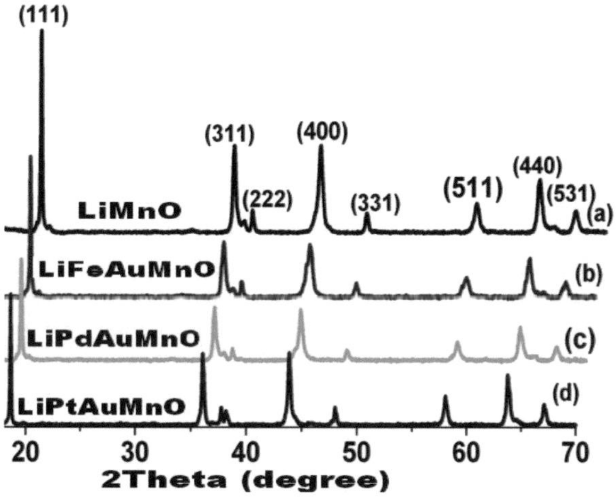

Figure 4. XRD spectra of $LiMn_2O_4$ and $Li(M)_{0.02}Mn_{1.98}O_4$ cathode materials.

Complementary to XRD, NMR is a useful tool to obtain information on the chemical and structural local environment of the nucleus under observation (^7Li). Any relevant structural change will affect the NMR spectrum, such as the change in bonding geometry, bonding distance, or nuclear charge [25]. The NMR spectra of $LiMn_2O_4$ and $Li(M)_{0.02}Mn_{1.98}O_4$ cathode materials are shown in **Figure 5**. At 16 kHz spinning speed, a high-resolution MAS NMR spectrum was obtained. The ^7Li-NMR spectrum of $LiMn_2O_4$ shows that there are two different types of lithium sites. The main peak at about 530 ppm is assigned to Li^+ in the tetrahedral 8a site of the spinel structure; the small resonances at about 580 and 615 ppm ($Li(M)_{0.02}Mn_{1.98}O_4$) are ascribed to Li^+ residing near the higher oxidation state manganese ions (Mn^{4+}) [26]. The lithium moves onto the tetrahedral site due to M_x atom propensity for the octahedral sites.

Therefore, the isotropic resonance at 511 ppm for $LiMn_2O_4$ is assigned to Li^+ ions in the tetrahedral 8a site, whereas the isotropic resonances at ~680 and 835 ppm for $Li(M)_{0.02}Mn_{1.98}O_4$ samples are assigned to lithium present in the proximity of higher oxidation state manganese ions (Mn^{4+}). These results suggest that the samples are pure and structural integrity is maintained.

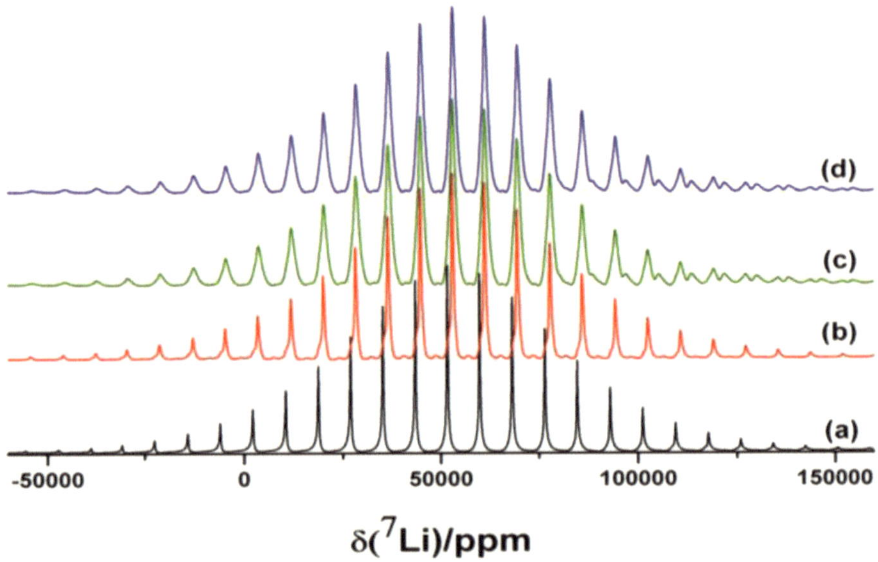

Figure 5. ^7Li NMR spectra of $LiMn_2O_4$ (a) and $Li(M)_{0.02}Mn_{1.98}O_4$ (b–d) at 16 kHz.

3.1. Electrochemical analysis

The ability of the metal alloy to enhance $LiMn_2O_4$ redox properties has been confirmed. The galvanostatic curve of $LiMn_2O_4$ (**Figure 6a**) is compared to that of $Li(M)_{0.02}Mn_{1.98}O_4$ cathodes (**Figure 6b–d**) at 0.1 mV s^{-1}. The pristine $LiMn_2O_4$ shows several peaks, which are ascribed to the $Mn^{3+/4+}$ redox reaction and constituents in the cathode material. Similar peak pairs are also found in the profiles of all functionalized samples with some shifts in potential and current. This demonstrates that coating does not change the electrochemical behavior but only the kinetics of $LiMn_2O_4$. The pair of peaks observed at ~3.2 V in the profiles of $Li(M)_{0.02}Mn_{1.98}O_4$ is well defined at 0.1 mV s^{-1}. This peak pair has been considered as the signature of $M^{3+/4+}$ redox reactions or its partial influence to a lesser extent in alloy functionalized spinel cathode material. Therefore, peak A2 is less pronounced in the pristine $LiMn_2O_4$ curve. $Li(M)_{0.02}Mn_{1.98}O_4$ has reduced anodic/cathodic potential differences compared to $LiMn_2O_4$, which is suggestive of smaller polarization due to the faster insertion/extraction of Li^+ ions in the spinel structure. This indicates that effective electron hopping between Mn^{3+} and Mn^{4+} lowers the potential barrier for Li^+ diffusion within the spinel lattice [27]. The faster lithium mobility enhances high rate performances [28]. This was corroborated by the high rate cycliability study.

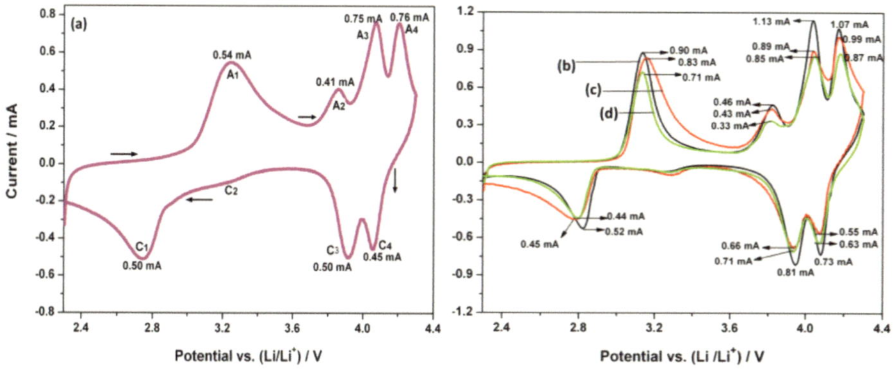

Figure 6. Cyclic voltammograms of $LiMn_2O_4$ (a), $LiFeAu_{0.02}Mn_{1.98}O_4$ (b), $LiPdAu_{0.02}Mn_{1.98}O_4$ (c), and $LiPtAu_{0.02}Mn_{1.98}O_4$ (d) cycled at 0.1 mV s^{-1}.

Figure 7 shows the cycling performance of $LiMn_2O_4$ and $Li(M)_{0.02}Mn_{1.98}O_4$ [M=FeAu (green), PdAu (black), and PtAu (red)] at 0.1 and 10 C discharge current rates. All the coated samples have stable discharge capacities at 10 C but show gradual decrease in capacity at 0.1 C. At 10 C, the discharge capacity decreases early for $LiFeAu_{0.02}Mn_{1.98}O_4$, with only 54% capacity retention. This suggests that ferric ions are probably involved in the reduction process leading to Fe^{2+}. When the cycling potential range is enlarged to overcome electrode kinetic limitations, the $Li(M)_{0.02}Mn_{1.98}O_4$ samples delivered relevant discharge capacities (70, 80, and 90 mAh g^{-1}) at high current density compared to that of $LiMn_2O_4$ (45 mAh g^{-1}). The $LiPtAu_{0.02}Mn_{1.98}O_4$ material exhibits the best capacity retention at 10 C with 88.7 mAh g^{-1} at the 50th cycle (99% capacity retention).

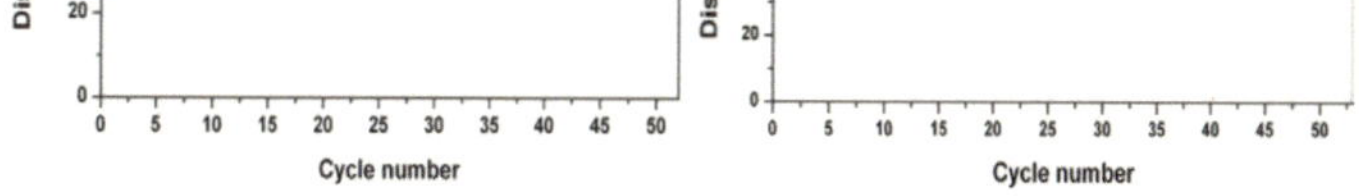

Figure 7. Cyclability of (a) $LiMn_2O_4$ and (b) $LiFeAu_{0.02}Mn_{1.98}O_4$ (green), $LiPdAu_{0.02}Mn_{1.98}O_4$ (black), and $LiPtAu_{0.02}Mn_{1.98}O_4$ (red) at 0.1 and 10 C.

The coulombic energy efficiency, which is the fraction of the electrical charge stored during charging that is recoverable during the discharge [29] of $LiMn_2O_4$, $LiFeAu_{0.02}Mn_{1.98}O_4$, $LiPdAu_{0.02}Mn_{1.98}O_4$, and $LiPtAu_{0.02}Mn_{1.98}O_4$ at 0.1 and 10 C discharge current rates, was computed using Equation (1):

$$CE = \frac{\int_0^t I_d dt}{\int_0^t I_c dt} \quad (1)$$

where I_d is the discharging current and I_c is the charging current, which are constant in most operations. With the increase in current density, the efficiency decreased to some extent. However, even at a current of 10 C, a maximum efficiency of 99% was obtained, confirming the excellent performance of the $Li(M)_{0.02}Mn_{1.98}O_4$ modified cathodes. The losses that reduce coulombic efficiency are primarily due to the loss in charge due to secondary reaction, such as the electrolysis of water or other redox reactions in the battery. The fading rate of $LiMn_2O_4$ at 0.1 C was reduced by ~15% by surface modification. At higher C rate, however, the fading rate was less apparent. The improved cycling performance is attributed to the stabilization of the spinel structure by metal cations [30]. An electrochemical impedance study [31] further corroborated the excellent high rate performance of the modified cathode. **Figure 8** shows that the total impedance observed for $Li(M)_{0.02}Mn_{1.98}O_4$ was moderately lower than that of $LiMn_2O_4$. The values obtained fitting to an equivalent circuit are shown in **Table 1**, confirming the integration of a conductive surface material.

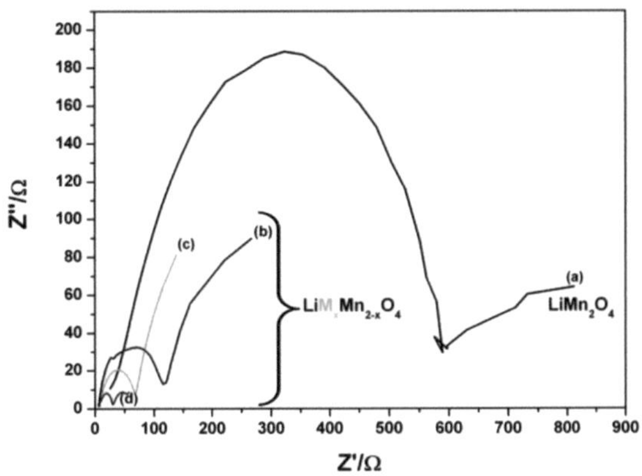

Figure 8. Nyquist plot of (a) $LiMn_2O_4$ and $Li(M)_{0.02}Mn_{1.98}O_4$ (b–d).

	$LiMn_2O_4$(a)	$LiFeAu_xMn_{2-x}O_4$(b)	$LiPdAu_xMn_{2-x}O_4$(c)	$LiPtAu_xMn_{2-x}O_4$(d)
R_{ct} (Ω)	140	93.25	80.86	30.85
f_Φ (kHz)	0.93	4.47	5.75	6.47

Table 1. Electrochemical impedance spectroscopy (EIS) characterization data of $LiMn_2O_4$ and alloy functionalized cathodes.

The kinetic index (calculated from f_Φ), explained by the time constant (τ), where τ is inversely proportional to the frequency and solution resistance as expressed by Equation (2), was faster

http://dx.doi.org/10.5772/62319

at the modified electrodes (2.41×10^{-4} s rad^{-1}) compared to the pure material (3.41×10^{-4} s rad^{-1}). This confirmed both the conductive and catalytic influences of the nanoalloy [32].

$$\tau = \frac{1}{4\pi f_{\phi}} \sqrt{1 + \frac{R}{R_s}} \tag{2}$$

Moreover, all $Li(M)_{0.02}Mn_{1.98}O_4$ samples showed improved discharge current densities. These samples exhibited two plateaus in the discharge curves, due to the two-step reduction and oxidation process, which is a characteristic of lithium manganese oxide spinel [33]. This further supports the XRD finding showing that the spinel structure of $LiMn_2O_4$ is retained after surface modification. The initial discharge capacities of $Li(M)_{0.02}Mn_{1.98}O_4$ increase due to the decrease in Mn^{3+} content. This means that the spinel phase is stabilized by the surface modification, and the alloy does not block the lithium pathway. Instead, the larger interface area provides more Li^+ ions for diffusion; therefore, the specific capacity is increased [34,35].

4. Conclusion

A novel transition metal alloy (FeAu, PdAu, and PtAu), functionalized $LiMn_2O_4$, was successfully synthesized. Excellent reproducibility was obtained when electrodes were cycled at high discharge rates. The $Li(M)_{0.02}Mn_{1.98}O_4$ cathode materials were lattice defective and phase pure and exhibited improved rate capabilities and improved electrochemistry compared to the pristine $LiMn_2O_4$. These improvements were attributed to enhanced electronic conductivity, lithium diffusivity due to the reduction in particle size, and structural stability. Microscopic results revealed that the composite cathode materials had well-crystallized particles and more regular morphological structures with narrow size distributions, which effectively accommodated the structural transformations that occur during Li^+ insertion. Moreover, these materials showed to be uniquely durable and might be used extremely efficient, making it economically viable. The enhancement of the capacity retention and higher electrode coulombic efficiency of the modified materials were significant, especially at high C rate.

Author details

Natasha Ross* and Emmanuel I. Iwuoha

*Address all correspondence to: nross@uwc.ac.za

SensorLab, Department of Chemistry, University of Western Cape, Private Bag X17, Bellville , Cape Town, South Africa

References

[1] Lu K. Materials in Energy Conversion, Harvesting, and Storage (111888910X); 2014. ISBN: 978-1-118-88910-7. 448.

[2] Eriksson T. $LiMn_2O_4$ as a Li-Ion Battery Cathode; From Bulk to Electrolyte Interface. Eklunds hofs Grafiska AB, Uppsala; 2001. Available at: https://uu.diva-portal.org/smash/get/diva2:160906.

[3] Guo YG, Hu JS, Wan LJ. Nanostructured materials for electrochemical energy conversion and storage devices. Adv. Mater. 2008;20:2878.

[4] Hunter JC. Preparation of a new crystal form of manganese dioxide: λ-MnO_2. J. Solid State Chem. 1981;39:142.

[5] Pasquier BA, Courjal AD, Larcher P, Amatucci D, Gerand G. Mechanism for limited 55°C storage performance of $Li_{1.05}Mn_{1.95}O_4$ electrodes. J. Electrochem Soc. 1999;146:428.

[6] Ohzuku T, Makimura Y. Layered lithium insertion material of $LiCo_{1/3}Ni_{1/3}Mn_{1/3}O_2$ for lithium-ion batteries. Chem. Lett. 2001;30:642.

[7] Mui SC, Trapa PE, Huang B, Soo PP, Lozow MI, Wang TC, Cohen RE, Mansour AN, Mukerjee S, Mayes AM, Sadoway DR. Block copolymer-templated nanocomposite electrodes for rechargeable lithium batteries. J. Electrochem. Soc. 2002;149:A1610.

[8] Spong AD, Vitins G, Owen JR. A solution-precursor synthesis of carbon-coated $LiFePO_4$ for Li-ion cells. J. Electrochem. Soc. 2005;152:A2376.

[9] Arico AS, Scrosati B, Tarascon JM, Van Schalkwijk W. Nanostructured materials for advanced energy conversion and storage devices. Nat. Mater. 2005;4:366.

[10] Xiong H, Guo J, Amemiya S. Probing heterogeneous electron transfer at an unbiased conductor by scanning electrochemical microscopy in the feedback mode. Anal. Chem. 2007;79:2735–2744.

[11] Haruta M. When gold is not noble: catalysis by nanoparticles. Chem. Rec. 2003;3:75.

[12] Liu J, Manthiram A. Materials challenges and opportunities of lithium ion batteries. Chem. Mater. 2009;21:1695.

[13] Jellinek J. Nanoalloys: from theory to application. Faraday Discuss. 2008;138:1.

[14] Liang AS, Caruso F. Gold nanoparticle-based core-shell and hallow spheres and ordered assemblies thereof. Chem. Mater. 2003;15:3176.

[15] Ferrando JR, Johnston RL. Nanoalloys: from theory to applications of alloy clusters and nanoparticles. Chem. Rev. 2008;108:845.

[16] Burda C, Narayanan R, El-Sayed MA. Chemistry and properties of nanocrystals of different shapes. Chem. Rev. 2005;105:1025.

[17] Zhao D, Xu B-Q. Enhancement of Pt utilization in electrocatalysts by using gold nanoparticles. Angew. Chem. 2006;30:5077.

[18] Shah A, Rahman L-u, Qureshi R, Rehman Z-u. Synthesis, characterization and applications of bimetallic (Au-Ag, Au-Pt, Au-Ru) alloy nanoparticles. Rev. Adv. Mater. Sci. 2012;30:133.

[19] Liu Q, Wang S, Tan H, Yang Z, Zeng J. Preparation and doping mode of doped $LiMn_2O_4$ for Li-ion batteries. Energies 2013;6:1718.

[20] Li X, Xu Y, Wang C. Novel approach to preparation of $LiMn_2O_4$ core/$LiNiXMn_{2-x}O_4$ shell composite. Appl. Surf. Sci. 2009;255:5651.

[21] Matsuda K, Taniguchi I, Ronbunshu KK. Particle properties of $LiMn_2O_4$ fabricated by ultrasonic spray. Pyrolysis method. J. Power Sources 2003;29:232.

[22] Otero F, Rodriguez J. Reinterpretation of polypyrrole electrochemistry after consideration of conformational relaxation processes. J. Phys. Chem. B 1997;101:3688.

[23] Hunter JC. Preparation of a new crystal form of manganese dioxide. J. Solid State Chem. 1981;39:142.

[24] Zhang D, Popov BN, White RE. Electrochemical investigation of $CrO_{2.65}$ doped $LiMn_2O_4$ as a cathode material for lithium-ion batteries. J. Power Sources 1998;76:81.

[25] Lee YJ, Grey CP. ^{6}Li magic-angle spinning (MAS) NMR study of electron correlations, magnetic ordering, and stability of lithium manganese (III) oxides. Chem. Mater. 2000;12:3871.

[26] Kumar VG, Gnanaraj JS, Ben-David S, Pickup DM, Van-Eck ERH, Gedanken A, Aurbach D. An aqueous reduction method to synthesize spinel-$LiMn_2O_4$ nanoparticles as a cathode material for rechargeable lithium-ion batteries. Chem. Mater. 2003;15:4211.

[27] Zhang X, Zheng H, Battaglia V, Axelbaum RL. Electrochemical performance of spinel $LiMn_2O_4$ cathode materials made by flame-assisted spray technology. J. Power Sources 2011;196:3640.

[28] Atanasov M, Barras J-L, Benco L, Daul C. Theoretical studies on the electronic properties and the chemical bonding of transition metal complexes using DFT and ligand field theory. J. Am. Chem. Soc. 2000;122:4718.

[29] Braun PV, Cho J, Pikul JH, King WP, Zhang H. High power rechargeable batteries. Curr. Opin. Solid State Matter 2012;16:186–198.

[30] Tang D, Ben L, Sun Y, Chen B, Yang Z, Gu L, Huang X. Electrochemical behavior and surface structural change of $LiMn_2O_4$ charged to 5.1 V. J. Mater. Chem. A 2014;2:14519–14527.

[31] Chen J. Recent progress in advanced materials for lithium ion batteries. Materials 2013;6:156–183.

[32] Mendelson MI. Learning Bio-Micro-Nanotechnology. CRC Press; 2012.

[33] Thirunakaran R, Gopukumar S, Rajalakshmi R. Cerium and zinc: dual-doped $LiMn_2O_4$ spinels as cathode material for use in lithium rechargeable batteries. J. Power Sources 2009;187:565.

[34] Wang X, Tanaike O, Kodama M, Hatori H. High rate capability of the Mg-doped Li-Mn-O spinel prepared via coprecipitated precursor. J. Power Sources 2007;168:282–287.

[35] Guohua Li, Ikuta H, Uchida T, Wakihara M. The spinel phases $LiM_yMn_{2-y}O_4$ (M=Co, Cr, Ni) as the cathode for rechargeable lithium batteries. J. Electrochem. Soc. 1996;143:178–182.

Chapter 3

Cathode Materials for Lithium Sulfur Batteries: Design, Synthesis, and Electrochemical Performance

Lianfeng Duan, Feifei Zhang and Limin Wang

Additional information is available at the end of the chapter

http://dx.doi.org/10.5772/62439

Abstract

With the rapid development of electronic devices, portable electronics, and electric vehicles, the energy density and cycle life of LIBs are insufficient for the demands. Based on the reaction mechanisms, lithium-sulfur (Li-S) batteries have a high specific capacity of 1672 mAh/g, with a theoretical energy density up to 2600 Wh/Kg. However, the sulfur cannot serve as cathode individually because of its isolation nature and soluble compounds, which necessitates a second component as a conducting matrix and sulfur host. Thus, sulfur cathodes have diversified through microstructure designing with various materials, including inorganic compounds, polymers, carbon materials, and their hybrids, which should be satisfied several essential requirements, such as high stable incorporation with sulfur, high electrical conductivity of electrode materials, and loose framework to suffer the volume expansion of cathode during charge-discharge process. These investigations may provide the effective routes to prepare different new cathode materials with unique structures and morphologies for Li-S batteries, which improve cycling stability, coulombic efficiency, and rate capacity of the electrode at higher current density.

Keywords: Cathode Materials, Li-S Batteries, Composites, Electrochemical Property, Electrochemical Performance

1. Introduction

With the fantastic development of electronic devices, ranging from portable electronics to electric vehicles, widespread concern for high-power and high-energy density rechargeable batteries has been paid in the new century [1, 2]. In particular, lithium-ion batteries (LIBs) are promising energy storage devices and achieved remarkable success in portable electronics market during the past two decades [3]. However, the traditional cathode materials based on the lithiated transition-metal oxide and phosphate have relative low energy densities (theoretically, 350–400 Wh kg^{-1}, practically, 100–220 Wh kg^{-1}) and insufficient for the upcoming large-scale energy demands [4]. In addition, most LIBs are too expensive to be economical for large-scale applications and make the price of electric vehicles prohibitively expensive for the ordinary consumers [5]. However, the limited electrochemical stability window of the currently available liquid electrolytes makes it difficult to increase the cathode operating voltage beyond 4.3 V. Also, the capacities of the insertion-oxide cathodes have reached a limit of 250 mA h g^{-1}. On the other hand, the capacity of the graphite anode is also limited to 370 mA h g^{-1}. Therefore, it is necessary to explore novel rechargeable systems based on new chemical theory, such as lithium-sulfur (Li-S) batteries, Li-air batteries and sodium-sulfur batteries, or new electrode materials with high energy densities, like multi-electron reaction materials and high voltage materials, to meet the requirements of sustainable energy fields.

Li-S batteries based on elemental sulfur as cathode and metal lithium as anode have attracted considerable attention because of their high theoretical specific capacity of 1672 mA h g^{-1} and high theoretical energy density of 2600 Wh kg^{-1} [2]. Moreover, as a byproduct of petroleum refinement or direct extraction from sulphate minerals, sulfur with the merits of abundant resources and environment friendly are beneficial for sustainable development [6]. The history of Li-S batteries development dated back to 1962 when Herbet and Ulam first used sulfur as a cathode material in electric dry cells and storage batteries [7]. Later, Argonne national laboratory developed a high-temperature Li-S system in 1967 [8]. However, the safety problems of high-temperature system remained challenging and limited applications, until E. Peled's group reported the electrochemical characteristics of sulfur in organic electrolytes at room temperature [9, 10]. Since then, more and more researches focused on the developments of high energy density, long-life times and safety Li-S systems.

2. Principles and challenges of lithium-sulfur batteries

2.1. The charge-discharge mechanism of Li-S batteries

Unlike the intercalation-deintercalation mechanism of traditional rechargeable LIBs, a Li-S cell undergoes a reduction-oxidation reaction and starts with discharge due to the charged state of sulfur. During the discharge reaction, lithium metal is oxidized to give up lithium ions and electrons. The lithium ions move from negative electrode to positive electrode through the internally electrolyte while the electrons circulate through the external electrical circuit, and thereby an electrical current is generated. Elemental sulfur at the positive electrode is reduced to lithium sulfide (Li_2S) by accepting the lithium ions and electrons [1]. The reverse reactions will occur during charge process. The discharge reactions can be described as follows:

Negative electode:

http://dx.doi.org/10.5772/62439

$$2Li \rightarrow 2Li^+ + 2e^-$$

Positive electrode:

$$S + 2Li^+ + 2e^- \rightarrow Li_2S$$

Overall discharge reaction:

$$2Li^+ S \rightarrow Li_2S$$

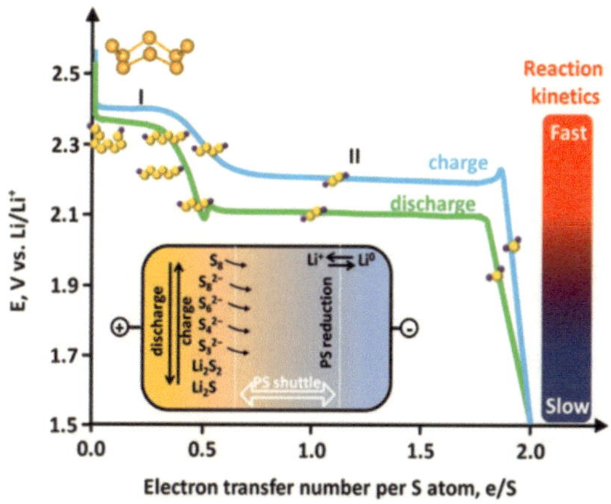

Figure 1. Electrochemistry of sulfur showing an ideal charge-discharge profile. Inset: polysulfide shuttle [13].

In fact, this reaction is a multi-step process which can be observed from the typical constant-current discharge curve of Li-S cell. As shown in **Figure 1**, there are two voltage plateaus in the first discharge curve. The higher voltage plateau at about 2.3 V is associated with the open ring reduction of the cyclic-octasulfur (S_8, most common stable form among sulfur allotropes) [14] to various long-chain (high-order) soluble lithium polysulfides (Li_2S_n, $4 \leq n \leq 8$), which proceeds through multiple steps [15, 16]:

$$S_8 + 2e^- \rightarrow S_8^{2-}$$

$$3S_8^{2-} + 2e^- \rightarrow 4S_6^{2-}$$

$$S_6^{2-} + 2e^- \rightarrow 3S_4^{2-}$$

The following plateau at about 2.1 V relates to the further reduction of these polysulfides to short-chain (low-order) insoluble lithium polysulfides (Li_2S_n, $n < 4$) and finally, lithium sulfide (Li_2S) [13].

$$S_4^{2-} + 4Li^+ + 2e^- \rightarrow 2Li_2S_2$$

$$Li_2S_2 + 2Li^+ + 2e^- \rightarrow 2Li_2S$$

During the following charge process, Li_2S is converted back into to intermediate lithium polysulfides, while the formation of S_8 does not usually occur [17, 18], the two charge voltage plateaus are normally overlapped with each other [1].

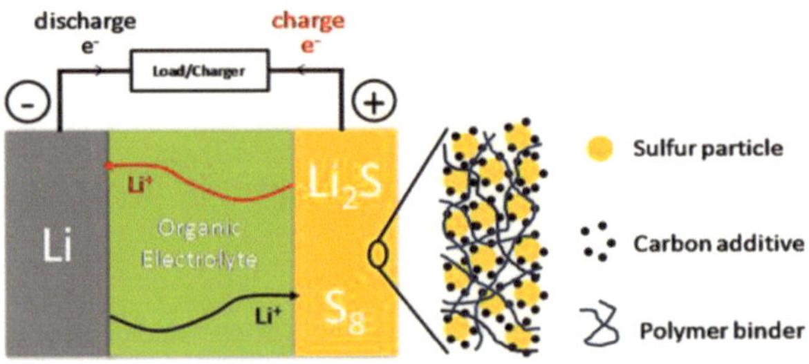

Figure 2. Structure scheme of Li-S battery [11].

2.2. The structure of Li-S batteries

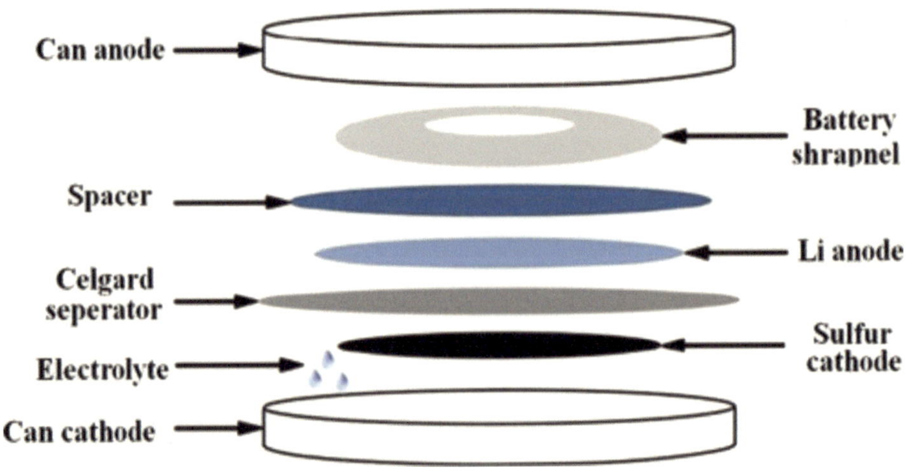

Figure 3. The schematic and assembling of CR2025 type button cell.

A Li-S cell is an electrochemical storage device through which electrical energy can be stored in sulfur electrodes. As shown in **Figure 2**, a conventional Li-S cell is composed of sulfur composite as the cathode, lithium metal as the anode, and separated an organic lithium solution as the electrolyte [12]. Because of the highly electrically and ionically insulating nature of sulfur, it must be kept at immediate contact with conductive additive to enable a reversible electrochemical reaction, then the cathode contains elemental sulfur, conductive carbon, and polymer binder. **Figure 3** is the assembling schematic of Li-S cell (CR2025 type button).

2.3. The challenges of sulfur cathode

Although the Li-S batteries exerted considerable superiorities when considering the high energy density and low cost, there are still many challenges of cathode materials associated with the practical applications.

Firstly, the cathode materials must be added with plentiful conductive agent to increase the electrical and ionic conductivity of cathode due to the insulating nature of sulfur (5×10^{-30} S cm^{-1}) [17], which resulting in low specific energy density of the whole cell. In addition, the last step of converting Li_2S_2 to Li_2S is difficult and is impeded by slow solid-state diffusion and poor electrical conductivity of Li_2S (3.6×10^{-7} S cm^{-1}) [15, 19]. Therefore, the active material utilization is low deriving from the incomplete discharge.

Secondly, the volume variation of sulfur particles during the charge and discharge process is large. Sulfur has a density of 2.03 g cm^{-3}, while Li_2S is lighter with 1.66 g cm^{-3}. As a result, the volume expansion within cathode when sulfur is fully converted to Li_2S is as large as 80% [13]. This volume change leads to serious cathode pulverization and thus fast capacity decay.

Thirdly, the polysulfide species (S_8^{2-}, S_6^{2-}, and S_4^{2-}) which are produced in the charge-discharge process are easily soluble in the electrolyte and diffuse freely from the cathode to anode driven by a concentration gradient (**Figure 2**, inset). Then, they react with lithium anode and reduced to short-chain polysulfides. The insoluble Li_2S_2/Li_2S deposits on the lithium anode can react with polysulfide ions again, yielding soluble medium-chain ions, which could diffuse back to cathode where they are reoxidized [20]. The mobile polysulfides therefore circulate between cathode and anode, driving the charging time of the cell towards infinity. This phenomenon, which is the main reason caused low coulombic efficiency and self-discharge of the cell, is so-called "polysulphide-shuttle" [21]. Additionally, these soluble species can be reduced to Li_2S insulating layer at the lithium anode surface, passivating the anode and leading to material loss and impedance increase. Moreover, the insoluble and electrically insulating Li_2S_2/Li_2S can be formed and precipitated out at the cathode surface. Once a thin Li_2S layer completely covers the whole cathode, further lithiation will be largely impeded and resulting in the low active utilization and coulombic efficiency. With the cycling going on, the dissolution and precipitation process will change the morphology of the cathode each cycle, which induces serious capacity fading and degrades the cycle life.

Excellent sulfur cathodes for high capacity and great cycling stability should contain several desired features: (1) a closed structure to encapsulate polysulfides efficiently, improving sulfur-electrolyte contact, (2) good conductivities achieved, for example, by mixing conductive materials in the cathode, (3) a loose structure to provide sufficient space to accommodate sulfur volumetric expansion and preserve the morphology of electrodes during transportation of electrons and Li ions.

In this chapter, different nanosized and nanostructured sulfur-based cathodes based on morphologies and compositions of the electrode are presented, including (1) sulfur-impregnated core-shell hierarchical porous carbon (HPC) composites; (2) sulfur/graphene nanosheets (GNS) composite; (3) sulfur@rGO composite with a saccule-like structure; (4) C-S@PANI composite with polymer spherical network structured. With such systematic discussion, we aim to provide a better and comprehensive understanding in this area. Although the crystal structures have boosted remarkable progress in the electrochemical property as cathode for Li-S batteries, it is more important to design and synthesis special and novelty cathodes in order to promote cycling stability, improve excellent capacity density, and rate capability by simple and environmental friendly processes. With such systematic discussion, we aim to provide a better and comprehensive understanding in this area.

3. Sulfur-impregnated core-shell hierarchical porous carbon composites as the cathode material for Li-S batteries

In order to advance Li-S batteries, a variety of strategies have been carried out to address these problems, including the optimization of electrolyte or additives [22], the preparation of the sulfur-conductive polymer composites [23], sulfur-carbon composites [24–26], and the protection of lithium anode. Among these approaches, fabricating sulfur/carbon composite cathodes with sulfur embedded within conductive carbon frameworks has been proven promising. In particular, carbon materials with porosity, high conductivity, large pore volume, and high surface area (such as carbon nanotubes, mesoporous and microporous carbon, hollow carbon, etc.) are proved to be effective and facile substrates for encapsulating polysulfides and improving sulfur-electrolyte contact [25–27].

3.1. The schematic representation of core-shell HPCs-S

Unique core-shell sphere with a radially aligned HPC shell is synthesized as the host of sulfur (**Figure 4a**). Both of the outer shell and inner core are highly porous framework, providing a fine carbon matrix for sulfur loading. Subsequently, the sulfur is impregnated by heating a mixture of the HPC composites and sulfur (HPCs-S) in a sealed inert atmosphere, forming core-shell HPCs-S-H composite (**Figure 4b**). The sulfur is not merely impregnated in the pores, but also deposited on the external of the spheres. Further treatment is carried out by dispersing the HPCs-S-H composite in the toluene to dissolve partial sulfur to obtain HPCs-S composite. The core-shell structure and open pore mouths of carbon sphere are maintained in HPCs-S composite. For this unique structure, sulfur is mainly distributing over the core area of HPCs, the shell of HPCs with void pores are serving as a retard against the dissolution of lithium polysulfides.

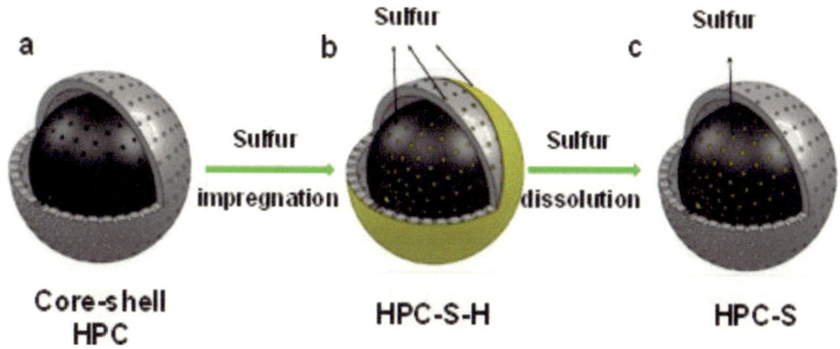

Figure 4. Schematic illustration of the procedure used to fabricate core-shell HPCs-S-H and HPCs-S composites [28].

3.2. Morphology and microstructure characterization of core-shell HPCs-S

Figure 5a shows the polymer-silica-surfactant (PSS) spheres are uniform solid spheres with perfectly smooth surface. After carbonized at 600°C and etched with HF solution, HPCs with pores on surface are obtained, and the dimension is 600~700 nm (**Figure 5b**). When impregnated with sulfur, a layer of sulfur particles deposited on the external of the spheres is clearly seen from SEM image of HPCs-S-H (**Figure 5c**). After the dissolution process, the sulfur aggregated on the surface vanishes and the porous morphology reappears (**Figure 5d**).

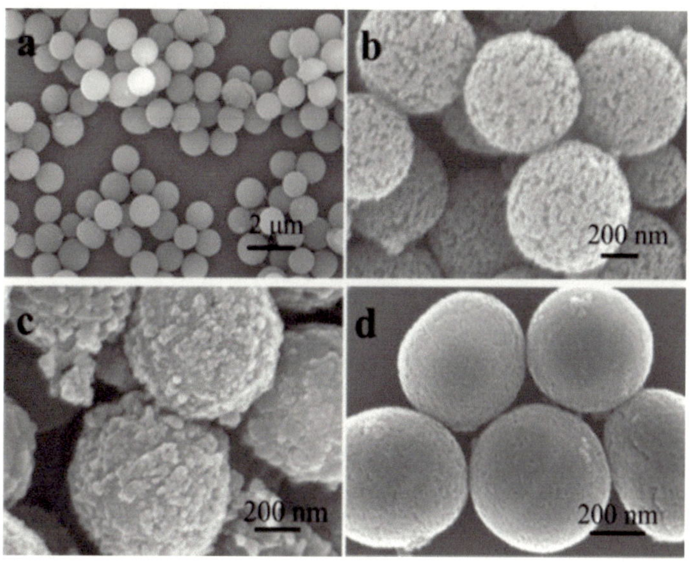

Figure 5. SEM images of the as-synthesized (a) PSS composite, core-shell (b) HPCs, (c) HPCs-S-H and (d) HPCs-S composite [28].

TEM images of HPCs are shown in **Figure 6**. The core-shell HPCs with highly porous structure can be obviously seen from **Figure 6a**. The dimension of spheres is in the range of 600–700 nm, and the thickness of shell is about 40 nm. TEM image in **Figure 6b, c** show that the distance of the gap between the shell and the core is about 8.3 nm. The outer shell is shucked and the inner mesopores are obviously observed. TEM image of the shell area which shows in **Figure 6d** exhibits that micropores and mesopores uniformly distribute through the shell area.

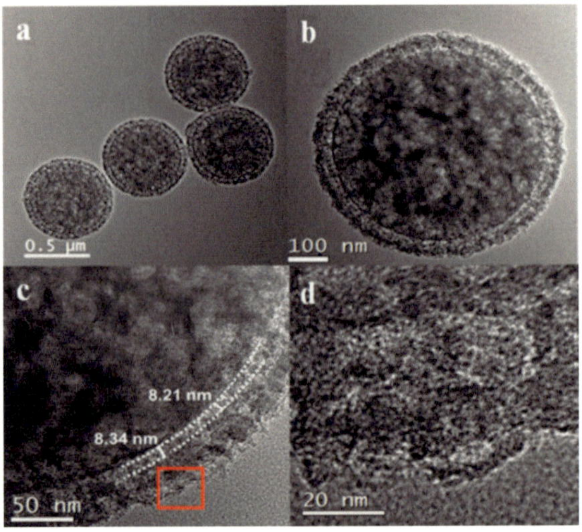

Figure 6. TEM images of the as-synthesized core-shell, (a) and (b) HPCs, (c) the gap between the shell and core, (d) the outer shell of an HPCs [28].

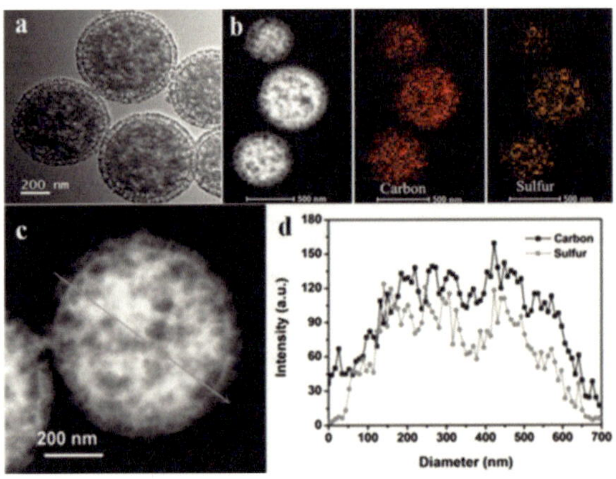

Figure 7. TEM images of the as-synthesized core-shell (a) HPCs-S composite, (b) magnified image of separate core-shel HPCs-S particles and the corresponding EDS elemental mapping of carbon and sulfur, (c) dark field TEM image of HPCs-S; (d) the linear EDX element distributions of sulfur and carbon along the arrow line of (c) [28].

Figure 7 is TEM images and elemental analysis of the as-synthesized core-shell HPCs-S composite. From this Figure, comparing with the HPCs, the core-shell structure and open pore mouths of carbon sphere in HPCs-S particles also can be obviously recognized. For HPCs-S, the carbon is distributing over the whole spheres, while the sulfur is mainly distributing over the core area of spheres, by analyzing the elemental mapping (**Figure 7b**). Moreover, linear EDX element distributions of sulfur and carbon along the HPCs-S as shown TEM image in **Figure 7c**. It exhibits the sulfur peak and carbon peak are both through the core area (**Figure 7d**), which indicates that most of extra sulfur is washed away during dissolution process.

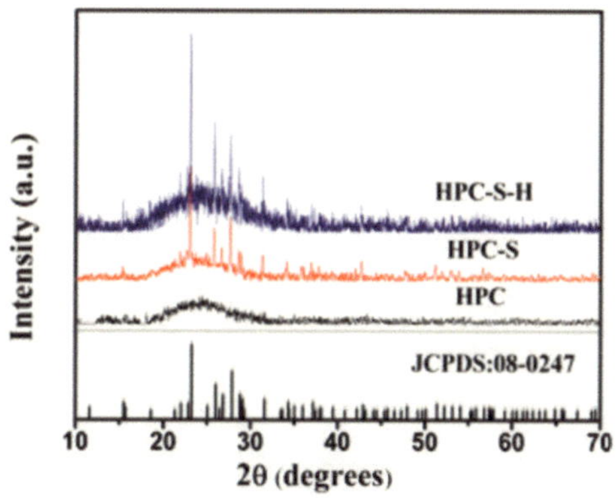

Figure 8. XRD patterns of the as-prepared core-shell HPCs, HPCs-S-H and HPCs-S composite [28].

Figure 8 is XRD patterns of HPCs, HPCs-S-H, and HPCs-S composites. The XRD pattern of HPCs shows a broad peak, which is assigned to the diffraction peak of amorphous carbon [29]. The patterns of HPCs-S-H and HPCs-S exhibit the broad porous carbon peak and sharp peaks for sulfur (JCPDS: 08-0247) [24]. However, it can be seen that the peak intensity of sulfur in the HPCs-S composite is weaker than the HPCs-S-H composite, because the sulfur is dissolving and aggregating on the surface of HPCs-S composite.

3.3. The specific surface areas and pore size distribution curve of core-shell composites

N_2 adsorption-desorption and the corresponding pore size distribution curve of HPCs, HPCs-S-H and HPCs-S composite are shown in **Figure 9**. The HPCs delivered a very high specific surface areas of 957.4 m^2 g^{-1} and a large total pore volume of 1.459 cm^3 g^{-1} (p/p_0 = 0.9703) (**Figure 9a, b**). Meanwhile, the average pore size of HPCs sample is 5.98 nm as calculated by the Barrett-Joyner-Halenda (BJH) method. From pore size distribution of the HPCs in **Figure 9b**, the micropores and macropores are 10.6% and 0.9% of the total pore volume, which could effectively hamper the dissolution of sulfur especially for long cycle process [30]. **Table 1** is summarized textural parameters of HPCs, HPCs-S-H, and HPCs-S composites. The mesopore and macropore volume of HPCs-S-H composite are also reduced sharply and micropores is disappear. It indicates that pores of the particles after loading sulfur into HPCs are almost filled with sulfur. And the micropores are fully filled with sulfur. Theoretically, the total pore volume (1.459 cm^3 g^{-1}) of HPCs can contain as much as 75.1 wt% sulfur (s = 2.07 g cm^{-3}) [31]. Thus the sulfur is not only impregnated in the pores, but also deposited on the external of the spheres. It corresponds to the morphology of HPCs-S-H in **Figure 5**. However, it is shown that the specific surface area and pore volume of HPCs-S composite are increased to 134.3 m^2 g^{-1} and 0.289 cm^3 g^{-1} after dissolution treatment (**Figure 9e, f**, and **Table 1**). When used as cathode material for Li-S battery, the void pores existed in the HPCs-S composite is beneficial to tolerate the volumetric expansion of sulfur after lithiation.

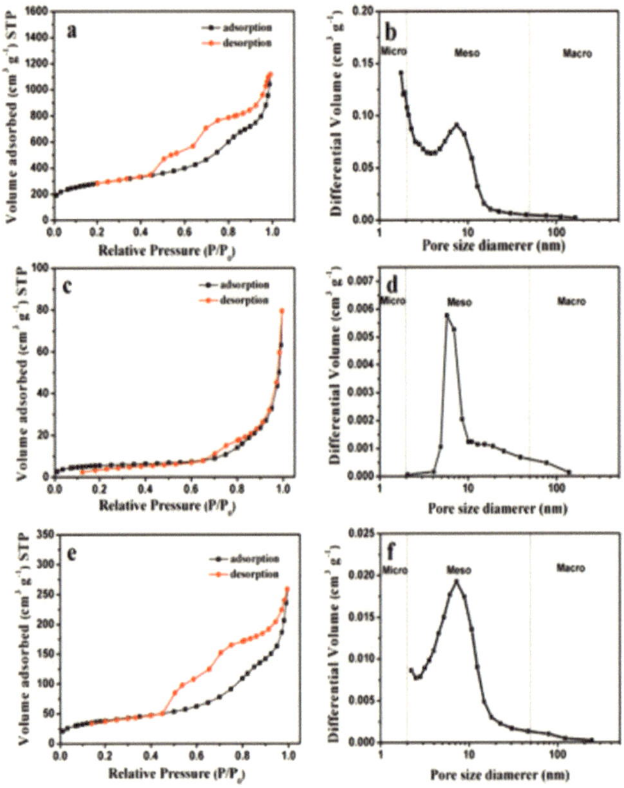

Figure 9. (a) N_2 adsorption-desorption and (b) the corresponding pore size distribution curve of HPCs composite; (c) N_2 adsorption-desorption and (d) the corresponding pore size distribution curve of HPCs-S-H composite; (e) N_2 adsorption-desorption and (f) the corresponding pore size distribution curve of HPCs-S composite [28].

Sample	BET surface area (m^2 g^{-1})	Total pore volume (cm^3 g^{-1})	Micropore volume (cm^3 g^{-1})	Mesopore volume (cm^3 g^{-1})	Macropore volume (cm^3 g^{-1})
HPCs	957.4	1.459	0.155	1.291	0.013
HPCs-S-H	18.6	0.067	0	0.0664	0.0006
HPCs-S	134.3	0.289	0	0.2873	0.0017

Table 1. Textural parameters of HPCs, HPCs-S-H and HPCs-S composites.

3.4. The electrochemical performance of core-shell HPCs-S

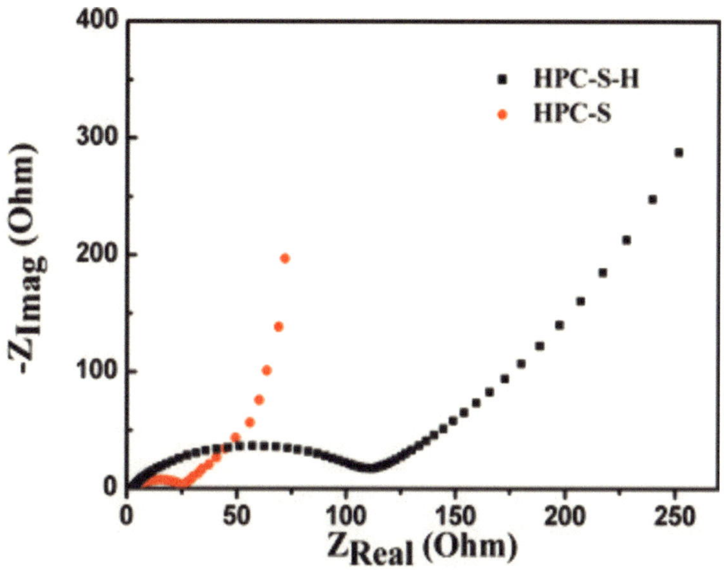

Figure 10. EIS measurements of the MHCs-S-H and MHCs-S electrodes [28].

Figure 10 shows the EIS results of HPCs-S-H and HPsC-S electrodes. In this figure, the resistance of HPCs-S-H electrode is almost four times that of the HPCs-S electrode because of the electrically insulating coating sulfur layer on HPCs-S-H composite isolate the conducting HPCs from the electrolyte, the resistant of HPCs-S-H composite electrode is higher [29, 32]. The CV curves of HPCs-S-H and HPCs-S electrodes in the potential range of 1.5–2.7 V is shown in **Figure 11a**. There are three reductive peaks in the first cathodic process of HPCs-S electrode, due to the multiple reductions of sulfur. The peak at 2.35 V corresponds to the reduction of S_8 to Li_2S_n ($2 \leq n \leq 8$). The other strong reduction of soluble lithium polysulfides to insoluble lower-order lithium sulfides (Li_2S_2 or Li_2S), because the intense cathodic peak is at 2.02 V. The reduction of high-order polysulfides to medium-order polysulfides, while the small peak is at 2.2 V [33]. There are two oxidation reaction peaks for the HPCs-S electrode. The first peak at 2.35 V is associated with the formation of Li_2S_n ($2 \leq n \leq 8$). The second peak at 2.45 V is the formation of elemental sulfur [33, 34]. **Figure 11b** is the charge-discharge voltage profiles of the HPCs-S-H and HPCs-S electrodes at 0.2 C (1 C = 1675 mA g^{-1}) in the voltage range of 1.5–2.7 V. The initial discharge capacity and a reversible specific charge capacity of the HPCs-S electrode are 1356 and 1397.9 mAh g^{-1}, which is much higher than the capacities of HPCs-S-H electrode (1080 and 1121.0 mAh g^{-1}). It demonstrates that utilization rate of sulfur for the HPCs-S electrode is much higher. And the electrode of the HPCs-S has sufficient contact with the electrolyte and the conducting material.

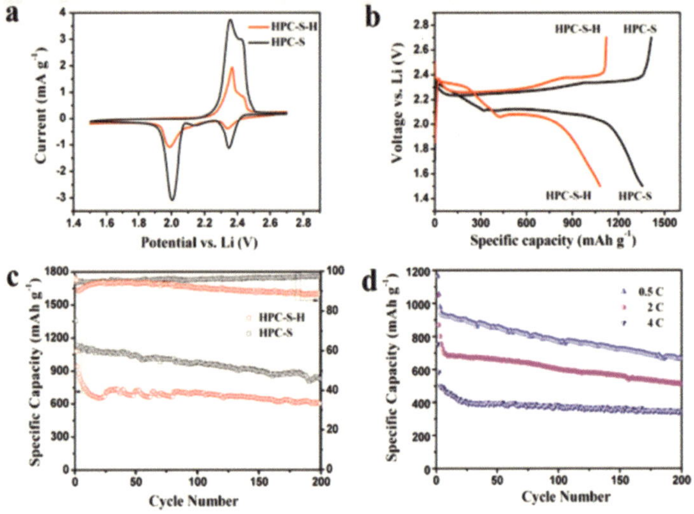

Figure 11. (a) Representative CVs of the HPCs-S-H and HPCs-S electrodes at a scan rate of 0.1 mV s^{-1} between 1.5 and 2.7 V vs. Li/Li^+. (b) Charge-discharge voltage profiles of HPCs-S-H and HPCs-S electrodes for the 1st cycle in the voltage range of 1.5–2.7 V at 0.2 C. (c) Capacity and Coulombic efficiency vs. cycle number of the HPCs-S-H and HPC-S electrodes at 0.2 C. (d) Rate capability of the HPCs-S electrodes [28]

The cycling abilities and the coulombic efficiencies of HPCs-S-H and HPCs-S electrodes at 0.2 C are presented in **Figure 11c**. The specific capacity of HPCs-S-H electrode declined sharply during the first 20 cycles, which is attributed to the dissolution of sulfur on the outer surface of core-shell HPCs [34]. After 200 cycles, the capacity of HPCs-S-H composite was 601.8 mAh g^{-1}. It indicates that the composite suffers a long-term loss of sulfur during cycling. After the exterior sulfur washed away, the reversible capacity of HPCs-S electrode was 830.5 mAh g^{-1} after 200 cycles, which showed much better cycling performance than HPCs-S-H composite. Because there are much more spare void pores and space on the shell of HPCs after dissolution process, it is favorable for electrolyte infiltration and electron transfer [28]. And also, it is beneficial to tolerate the volumetric expansion of sulfur after lithiation [31]. Meanwhile, the coulombic efficiency of HPCs-S electrode maintained about 95.0% during the cycling processes. It is also better than HPCs-S-H electrode. During charge-discharge process, the sulfur is reduced to dissolving polysulfide. Thus, the exterior sulfur layer and the polysulfides escaped from inner pores as the volumetric expansion are readily dissolved in the electrolyte and then lost, as a result, the capacity of HPCs-S-H with a continuous fading on the onward cycles. Furthermore, the soluble polysulfides result in the random precipitation of Li_2S_2 and Li_2S on the electrode, which changes the electrode morphology and increases the resistance of HPCs-S-H electrode [28]. In addition, rate capability behavior of the core-shell HPCs-S composites was detected. The discharge capacities remain 1163.9, 1049.0, and 753.2 mAh g^{-1} at 0.5, 2, and 4 C, separately (**Figure 11d**).

The significant electrochemical performance of the HPCs-S composite can be attributed to the hierarchical porous core-shell structure with high specific surface area and pore volume, which contain the substantial amount of sulfur, and suppress the diffusion of dissolved polysulfides.

And then, the dissolution treatment can prevent sulfur agglomerating on the surface and create void space on the shell of HPCs. It could enhance the transport of lithium ion and infiltration of electrolyte. At last, the outer shell with void pores can avoid effectively the diffusion of dissolved polysulfides and alleviate the stress caused by volume change during the charge-discharge process [28]. The HPCs-S composite with the hierarchical porous core-shell structure maintain the mechanical integrity and chemical stability during the cycling process, because various synergistic effects of the material characteristics and structure design, which could enhance the cycle performance and rate capability of the electrode materials.

4. Synthesized of the sulfur/graphene nanosheets composite

With a few atoms thickness, graphene inherently shows many advantages such as high surface area, excellent electrical conductivity, light weight, good flexibility, and superior mechanical strength, etc. Thus, it can be applied in electronics, engineering materials and is also considered as an ideal substrate for loading active materials like sulfur for electrochemical energy storage.

The application of graphene in Li-S battery is very promising due to its unique 2D structure, high conductivity and superior mechanical flexibility. The S distributed uniformly between the layers of graphene. This structure of carbon atoms is employed as a conductive and absorbing agent for the S cathode materials of rechargeable Li-S battery. Besides, the surface functional groups of graphene can be tuned flexibly to immobilize S/Li_2S_x on the graphene surface during the cycling process. In a recent report by Wang et al., S was melted and coated on the GNS, which shows an improved capacity [35]. Furthermore, a graphene-sulfur composite material was synthesized by synthesizing submicrometer sulfur particles coated with PEG containing surfactants and GNS. The composite showed high specific capacity with relatively good cycling stability as the cathode for Li-S batteries [25].

4.1. The fabrication of the sulfur/graphene nanosheets composite

The graphene oxide (GO) was made using a modified Hummers method [36]. The elemental S is mixed in GO solution at the weight ratio of 1:5, as the mixture designated S/GO solution. The S/GO was ultrasonicated for 30 min to get a uniform suspension and then lyophilized it. The mixture was held at 423 K for 4 h under argon gas protection to allow the melted elemental S to infiltrate into the layers of GNS [37]. Then, the temperature was increased to 573 K and held for 2 h, reduced GO to GNS.

4.2. Morphology and microstructure characterization of the sulfur/graphene nanosheets composite

The SEM images of element S, GNS, and S/GNS composites prepared through reduction are shown in **Figure 12**. In this figure, the morphology of bare S power is microsized particles (**Figure 12a**) and the morphology of graphene is nanosheets (**Figure 12b**), which was reduced by decomposition at 573 K for 4 h under argon gas protection. In **Figure 12c**, the microsized S particles melted and coated uniformly on the GNS sheets was prepared after heat treatment of S/GO. By the EDS mapping/imaging of the material, it was confirmed that element C and S. And the element S was coated on the GNS sheets uniformly from **Figure 12d, e**, and f.

Figure 12. SEM images of (a) elemental S powder, (b) GNS and (c) S/GNS composite. EDS characterization of S/GNS composite. (d) SEM image of GNS coated S particles. (e) EDS mapping of the carbon region shown in (d). (f) EDS mapping of the S region shown in (d) [37].

Figure 13 is the XRD patterns for the S powder and S/GNS composite. The elemental S exhibits several sharp peaks, indicating its good crystal state. However, the characteristic peaks of crystal S in the S/GNS composite with a low S content are not detectable. It demonstrate that S becomes amorphous and homogeneously distributed in the composite.

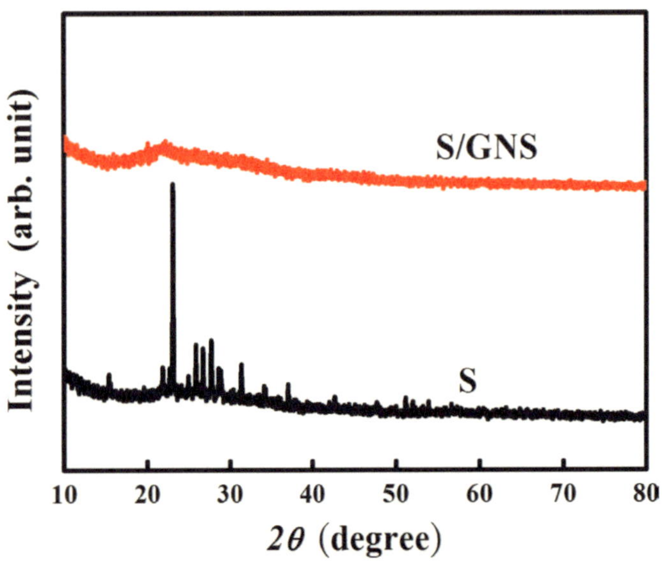

Figure 13. XRD patterns of elemental S and S/GNS composite [37].

4.3. The electrochemical performance of the sulfur/graphene nanosheets composite

Figure 14a shows the discharge curves of the S and S/GNS composite, which represent a typical characterization of Li-S batteries. The upper plateau shows the change from elemental S to the higher order Li polysulfides (Li_2S_n, $8 > n > 2$), and the lower plateau indicates the reduction of the higher order Li polysulfides to lower order Li sulfides (Li_2S_2, Li_2S). From this figure, the initial discharge specific capacity of S/GNS cathode is about 1598 mAh g^{-1}. In contrast, the initial discharge-specific capacity of the S cathode is about 1094 mAh g^{-1}. The cycling performances of the pure S and S/GNS composite electrode at the constant current density of 160 mAh g^{-1} S is shown in **Figure 14b**. The discharge specific capacity of the S/GNS composite electrode decreases from initial 1598 to 670 mAh g^{-1} after 80 cycles. Compared to the pure S electrode, the initial capacity and the cycling stability of the S/GNS electrode are both improved.

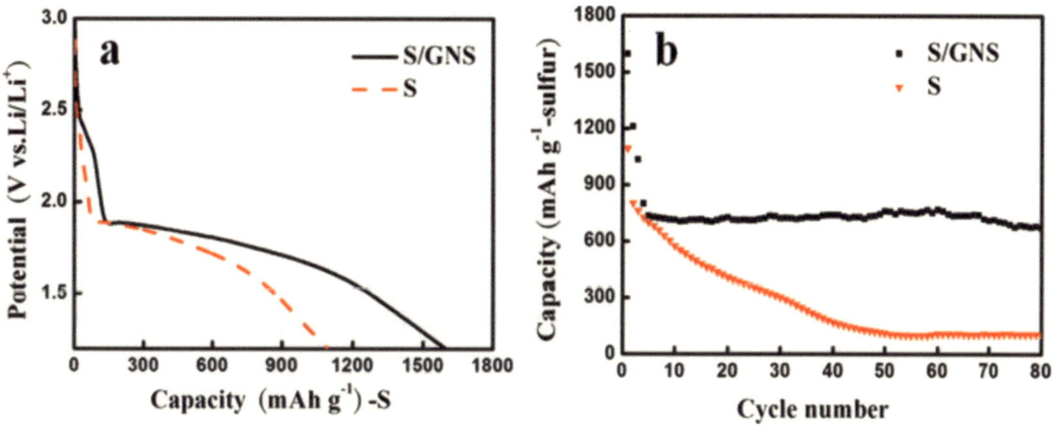

Figure 14. First discharge curves of the S and S/GNS composite electrodes and cycle performance for S and S/GNS composite electrodes [37].

Furthermore, EIS of the differences two electrodes is investigated in Nyquist profiles (**Figure 15**). It indicates that all the Nyquist plots of cathodes are composed by a semicircle at high frequencies relating to the contact resistance and charge transfer resistance, and a short inclined line in low-frequency regions due to the ion diffusion within the cathode [37]. It can be inferred that GNS can provide a better electron and ionic conductivity network, because the semicircle in S/GNS electrode is much smaller than the S electrode.

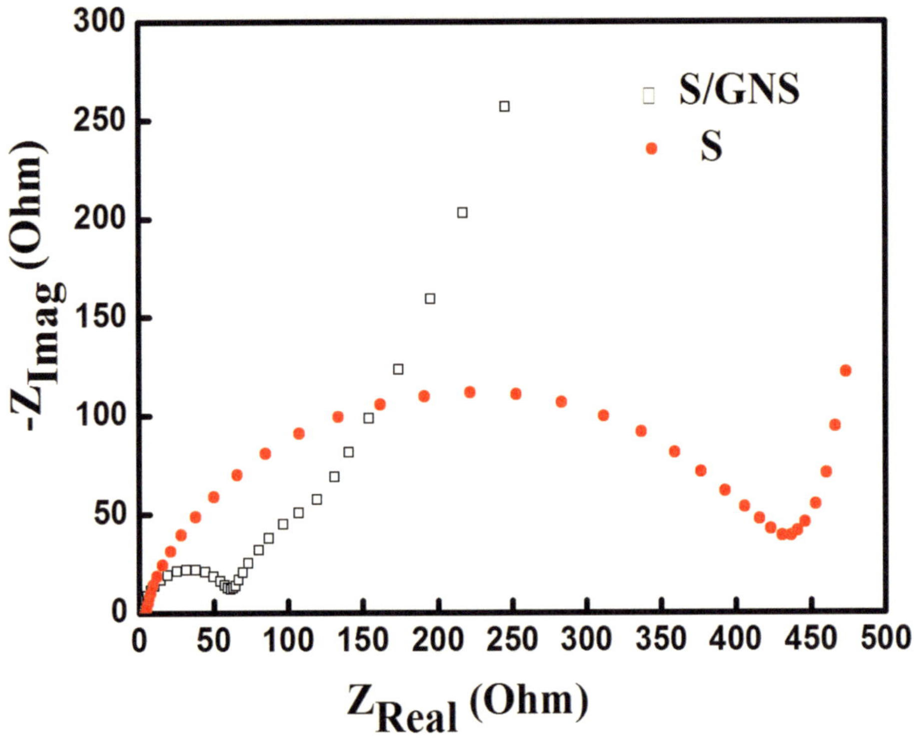

Figure 15. Impedance plots for electrodes of S and S-GNS composite [37].

5. Synthesized of a sulfur@rGO composite with a saccule-like structure

In contrast to graphene, GO is a compound of carbon, oxygen, and hydrogen in variable ratios and is rich in epoxide, carbonyl, hydroxyl, phenol, and organosulfate functional groups, whose structure and properties depend on the particular synthesis method and degree of oxidation [38]. The micro-sized sulfur nanoparticles enveloped by the rGO was synthesized by a scalable solution-based oxidation process. It had been achieved a high loading amount of sulfur (87%) embedded into the rGO matrix [39]. The synthesized GO is hydrophilic and easily dispersed in water, making it suitable for the synthesis of sulfur-graphene composites. In general, graphene-based materials are less expensive to produce than some other carbon nanomaterials such as carbon nanotubes, which is advantageous when they are applied as raw materials for the synthesis of graphene-based composite materials.

It has been reported that sulfur-graphene composite was synthesized by using several methods [40]. However, the 2D structure is not effective in containing the polysulfides, although the graphene could improve the conductivity of sulfur. In the recent reports, the GO-sulfur composite showed high specific capacity with relatively good cycling stability as the cathode for Li-S cells, which showed high specific capacity with relatively good cycling stability as the cathode for Li-S cells [41].

5.1. The schematic representation of sulfur@rGO composite with a saccule-like structure

The detailed synthetic procedures are shown in **Figure 16**. First, graphite oxide (GO) synthesized by a modified Hummers method is dispersed with water to form a homogeneous suspension (water-phase) under ultrasonic condition. Then the oil-phase containing sulfur in carbon disulfide (CS_2) is dropwise added into GO solution under ultrasonication. Followed by completely evaporation of CS_2 from the mixture, hydrazine hydrate ($N_2H_4 \cdot H_2O$) is then added to reduce GO to rGO. After stirring for 12 h at room temperature, the S@rGO composite is collected by consecutive centrifugation and water-washing cycles as well as lyophilization.

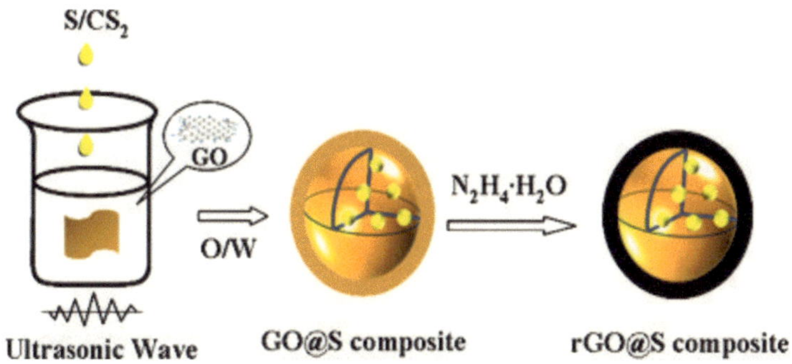

Figure 16. Schematic of synthesis steps for S@rGO composite [42].

5.2. Morphology and microstructure characterization of sulfur@rGO composite

The microstructures of the as-synthesized sample are characterized by SEM. The dimension of the composite particles before reduction is ca. 1 μm (**Figure 17a**). Meanwhile, the image shows GO sheets coat around the particles, indicating the feasibility of our strategy in preparing encapsulated structure using O/W system. Interestingly, after reduction with $N_2H_4 \cdot H_2O$, the size and surface morphology of the composite particles can be kept almost unchanged (**Figure 17c**). The morphology and structure of the interior sulfur were analyzed by TEM in **Figure 17d**. Unexpectedly, the interior sulfur encapsulated by rGO is composed of many smaller particles with the size ranging from 10 to 100 nm. When the oil droplets of CS_2 containing sulfur are first split into smaller droplets and surrounded by GO sheets under ultrasonic condition. Then, due to the volatile nature of CS_2, it gradually evaporated from the O/W system accompanied by forming the saccule-like structure and making the dissolving sulfur recrystallize into smaller particles [42]. It should be noted that the obtained saccule-like structure has enough space to accommodate volumetric expansion of sulfur during charge-discharge process.

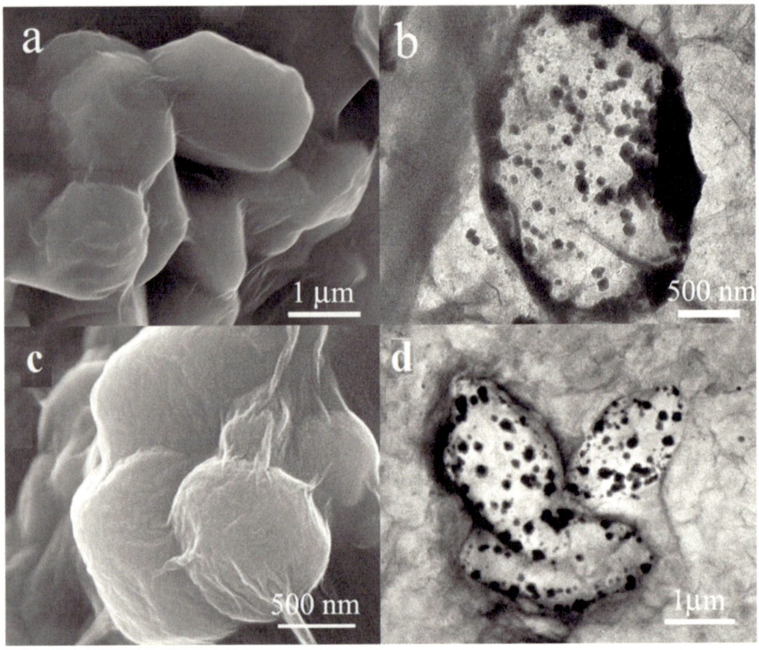

Figure 17. SEM and TEM images of S@GO and S@rGO composite [42].

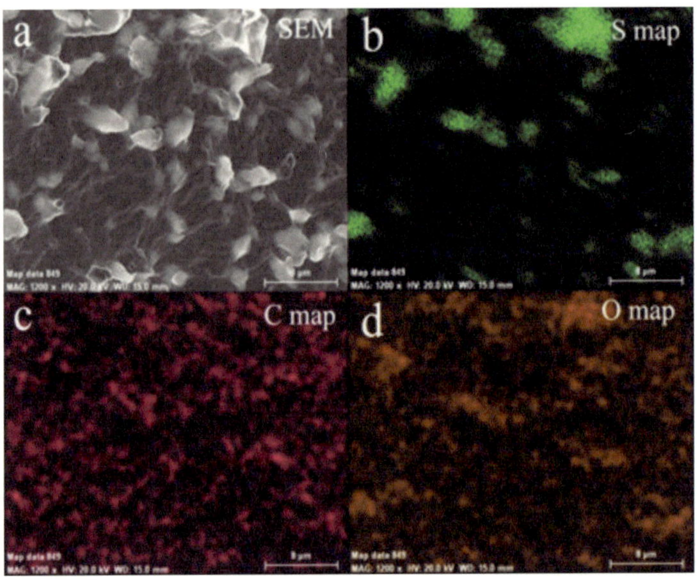

Figure 18. EDS characterization of S@rGO composite. (a) SEM image of rGO-coated sulfur particles. (b) EDS sulfur mapping of the region shown in (a). (c) EDS carbon mapping of the region shown in (a). (d) EDS oxygen sulfur mapping of the region shown in (a) [42].

In **Figure 18**, the structure and composition of the S@GO composite was verified by EDS mapping of the material. It is shown that sulfur is encapsulated by GO in the saccule-like structure. The relative change in the ratio of D to G peak intensity before and after the reduction in the Raman spectrum confirms the reduction of GO (**Figure 19a**) [28]. The XRD analyses are also conducted on the obtained S@rGO composite and elemental sulfur (**Figure 19b**). The reflections of the elemental sulfur showed two prominent peaks at 2θ = 23° and 28° corresponding to an F3d orthorhombic structure [43, 44]. As shown, the characteristic peaks in XRD pattern of the elemental sulfur and S@rGO composite remain the same, further indicating the existence of sulfur in the saccule-like structure. And then, the content of sulfur in the S@rGO is as high as 65 wt% by elementary analysis.

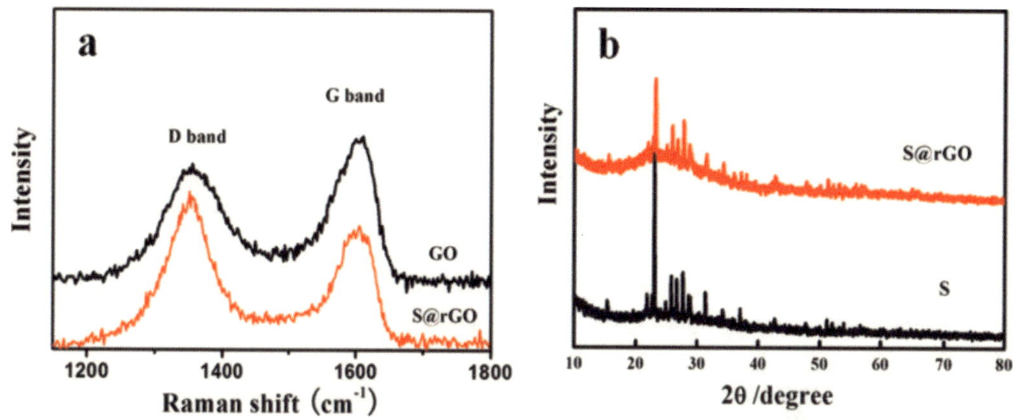

Figure 19. (a) Raman spectra of the GO solution and S@rGO composite, (b) XRD patterns of the element S and S@rGO composite [42].

5.3. The electrochemical performance of sulfur@rGO composite with a saccule-like structure

Figure 20a shows the first cycle charge and discharge curves of the sulfur@rGO composite at different high rates. The discharge capacity of composite is 724.5 mA h g^{-1} at a current rate of 1 C (1 C = 1675 mA g^{-1}). And also, the discharge capacity remains as high as 715.6, 708.6, and 697.5 mA h g^{-1} at 2, 3, and 4 C, respectively, which is much better than the S-rGO cathode. **Figure 21** is CV profile for indicating the electrochemical reactions of the composite cathode, which suggests a two-step reduction of sulfur. However, there is only one intensive oxidation peak at about 2.5 V, in the subsequent anodic scan, which indicates the conversion of Li_2S and polysulfides into elemental S.

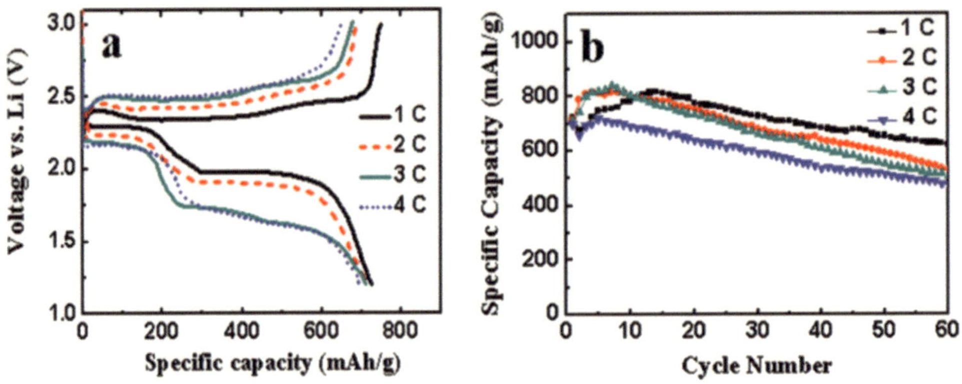

Figure 20. (a) First cycle charge/discharge voltage profiles of the S@rGO cathodes at various C rates. (b) Cycle performance of the S@rGO cathodes at various C rates [42].

The cycling performance of the S@rGO composite for the Li-S cell is shown in **Figure 20b**. The discharge capacity increased gradually with increasing cycle number during initial several cycles, because the S@rGO composite in the cathodes are incompletely soaking and penetrating with the electrolyte. After the 60th cycle, the cathode holds the reversible discharge specific capacity of 621.9 mA h g^{-1}, at a rate of 1 C. At the high rates of 2, 3, and 4 C, the capacities can remain 530.8, 505.8, and 478.7 mA h g^{-1}, respectively, and the 60th charge/discharge profiles are also shown the good performance of the cells.

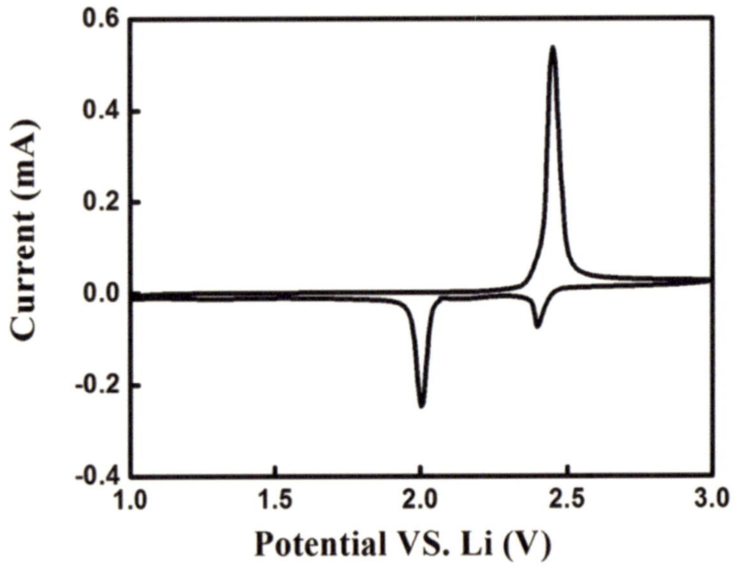

Figure 21. Cycling voltammetry curve of the first cycle at scan rate of 0.1 mV s^{-1} [42].

So, the good cycling stability of the cathode at high rates is owing to the excellent flexibility and mechanical strength of graphene, meanwhile, the unique saccule-like structure of the composite could provide enough space to accommodate stress and sulfur volumetric expansion during charge-discharge process [42]. Although the cycling stability and rate capacity of sulfur@rGO composite have been improved obviously, the problem of polysulfides dissolution has not been solved completely, leading to some capacity loss in the cycling process. So, the conductive polymer has been coated onto the surface of sulfur particles to improve the electronic conductivity.

6. Synthesized of the C-S@PANI composite with polymer spherical network structured

Focused on creating fine structures, the polymer/S-C composites utilize interactions between lithium ions in solution and functional group uniformly distributed along the chain backbone of a polymer precursor. The novelty structure could control the distribution of lithium sulfide in the host material similar to vulcanization methods [44]. In addition, some additives, especially $LiNO_3$, could enhance the coulombic efficiency via self-sacrifice [45]. Among the family of conductive polymers, PANI is attractive because of its associated electrical, electrochemical, optical properties and free volume, coupled with its excellent environmental stability [46, 47]. Cui's group also reported a nano-structured polymer-wrapped hollow sulfur nanospheres-based composite with enhanced specific capacity and cycling performance [48]. This approach is coating a conductive polymer onto the surface of sulfur particles to improve the electronic conductivity of cathodes and prevent the dissolution of polysulfides [42, 49]. These approaches improve the utilization of sulfur and cycle stability to different extents.

6.1. The schematic representation of the C-S@PANI composite with PSN

A new scheme is proposed to prepare composites based on lithium sulfide uniformly dispersed in Polymer Spherical Network by the in situ synthesis technique. The C-S@PANI composite with conductive PSN has been synthesized by a grafting method. **Figure 22** shows a schematic illustration of the synthesis process. Firstly, the suspension of oxidized carbon black (CBO) is prepared by dispersing stoichiometries Acetylene black (AC) in concentrated nitric acid (HNO_3). Secondly, hydrochloric acid solution (HCl) is prepared and mixed with CBO, adding aniline (ANi) under ice cooling (0–4°C) with magnetic stirring. Then the initiator (the molar amount of ammonium persulfate (APS) and ferric chloride ($FeCl_3$) is 1:0 to 1:1) is added dropwise while stirring the mixed suspension of CBO/ANi (speed < 0.5 mL/min). The solution is stirred, sealed, and reacted for t_1 (18 h) at 0–4°C, t_2 (24 h) at room temperature (20–30°C), ultrasonicated for 10 min, washed with acetone and deionized water, ultrasonicate after each wash and centrifuge to obtain the conductive polymer (CBO-PANI). And last, the sulfur of composite materials is filled by the liquid phase deposition method, first mixture (CBO-PANI) is added into HCl solution and stirred for 5 min [47]. Sodium thiosulfate ($Na_2S_2O_3{\cdot}5H_2O$) (rate v_2 < 1 mL/min) is added dropwise with magnetic stirring into the conductive frame (CBO-

PANI). After stirring, the solution is heated at 155°C for 18 h. Washed with deionized water and alcohol, with ultrasonication after each wash, the composite (C-S@PANI) is obtained through centrifugation [47].

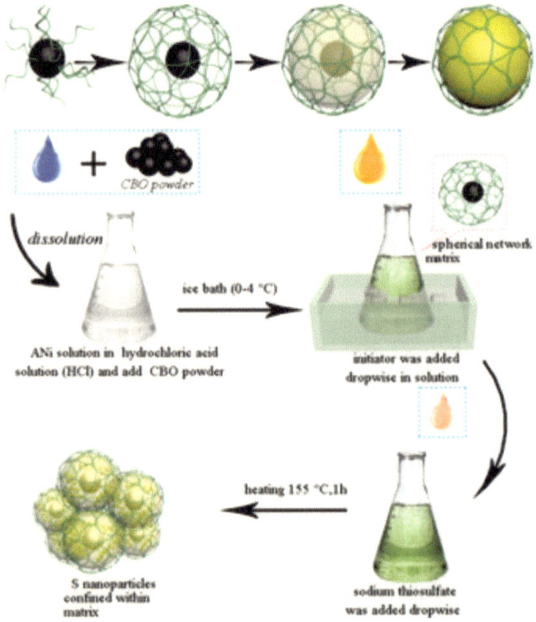

Figure 22. The design of PSN matrix (C-S@PANI composite) as the cathode materials for lithium-sulfur batteries (Green: PANI, yellow: sulfur, black: acetylene black), and schematic illustration of the synthesis process [47].

6.2. Morphology and microstructure characterization of the C-S@PANI composite with PSN

During the co-heating process, the molten sulfur can penetrate into the PANI spherical network. At the same time, the network shows strong absorption ability to receive elemental sulfur. **Figure 23** is the morphology of C-S@PANI composite. The typical matrix structure of the C-S@PANI composite with C-S particles reunited to form spherical particle with sizes of 80–100 nm is showed clearly in this figure. The whole particle congeries completely are enclosed and pinned by the PANI network matrix of nearly 20 nm in thickness. As shown in **Figure 23d**, the sulfur and carbon can maintain uniform distribution on a single spherical particle. It is also indicated that sulfur is encapsulated within the PANI matrix uniformly after heating. From morphology images of C-S@PANI composite, there are many pores, which can absorb polysulfide when the PANI layer connects to the C-S particles to form a micro-reactor. Furthermore, the pores provide channels for electrolyte infiltration and ion diffusion. At the same time, the pores provide much more spaces for volume expansion produced by the sulfur discharge products. On the other hand, chemical bond generated due to the chemical effect, the interaction between the functional group of conductive polymer surface, and the reactions products of the polysulfide [50], resulting in excellent sulfur wrapped.

http://dx.doi.org/10.5772/62439

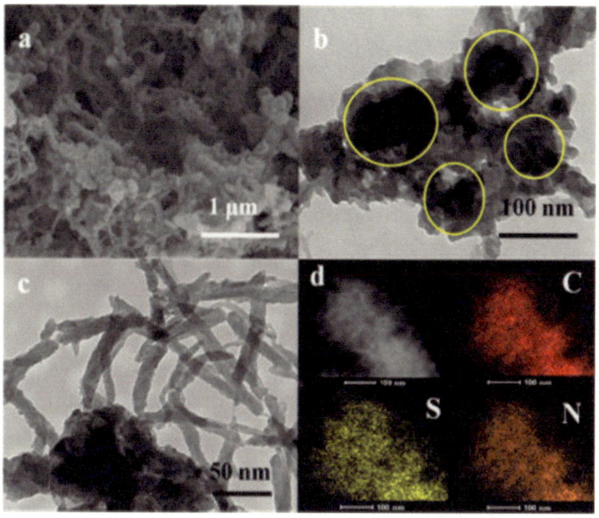

Figure 23. (a) SEM images and (b), (c) TEM images of C-S@PANI composites (d) EDX elemental mappings of carbon, sulfur and nitrogen [47].

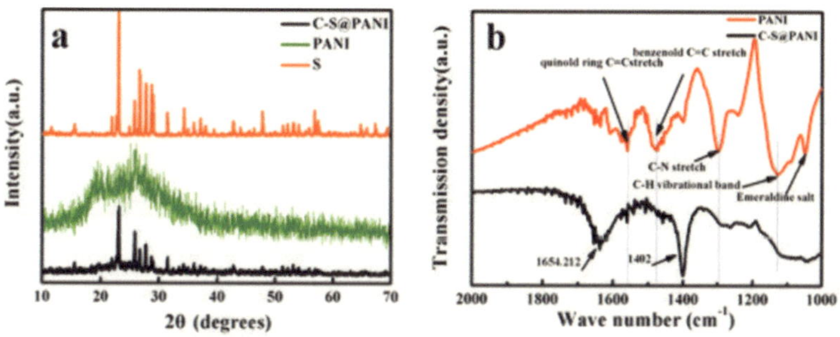

Figure 24. (a) XRD patterns of sublimed S, PANI and C-S@PANI composite 40% of sulfur, (b) FTIR spectra of the PANI and C-S@PANI composite [47].

The XRD analyses and FTIR spectra of the PANI and C-S@PANI composite after heat treatment are shown in **Figure 24**. In the XRD patterns, the C-S@PANI composite exhibits significantly lower peak intensities than the sulfur. It indicates the amorphous nature of the sulfur within the composite. FTIR measurement illustrates the presence of chemical bonds and functional groups between with PANI matrix and sulfur, in **Figure 24b**. The C=C stretching vibration at 1497 cm^{-1}, assigned to benzenoid rings shifts to lower wave numbers, which could be coming from the substitution of H atoms on benzenoid rings by S atoms [47]. The C-N stretching vibrational bands at 1298.4 cm^{-1} and the C-H vibrational band in the vicinity of 1128.7 cm^{-1} weakens significantly, further confirming the replacement of H atoms on aromatic rings by S atoms [47]. C-S@PANI composite with PSN provides strong physical and chemical confinement to the elemental sulfur and the resident polysulfide.

6.3. The electrochemical performance of the C-S@PANI composite with PSN

Figure 25 is the electrochemical performances of the C-S@PANI composite with PSN structure by galvanostatic charge/discharge cycling. In **Figure 26a**, there are two voltage plateaus at 2.3 and 2.0 V at 0.1 C (according to the theoretical capacity of sulfur is calculated 1 C=1672 mAh/g). The elemental sulfur capacity is calculated by subtracting the amount of carbon black and polyaniline from the (PSN/S). The high voltage plateau corresponds to the long polysulfide chains being produced during the first reduction step (2.4–2.2 V), such as S_8^{2-} and S_6^{2-}, as well as S_4^{2-} being produced during the second reduction step (2.15–2.1 V) [49]. The low-voltage plateau is attributed to the gradual decrease of the polysulfide chain lengths. Finally, short polysulfide species, such as S_3^{2-}, S_2^{2-}, and S^{2-}, are produced at the end of the reduction process [51]. The electrode exhibits a discharge and charge capacity of 608 and 628 mA h g^{-1} at 0.1 C for the first cycle. And the discharge and charge capacity increased to 1238 and 1453 mA h g $^{-1}$, after 20 cycles in **Figure 26a**. **Figure 26b** is cycling performance of C-S@PANI composites, which shows that C-S@PANI composite with PSN structure remain relatively stable at a value as high as 948 mA h g^{-1} after 200 cycles at 0.1 C. Meanwhile, the specific capacity of composite are 758 and 743 mA h g^{-1} after 200 cycles at 0.2 C and 0.5 C, respectively. In addition, as seen in **Figure 26c**, the discharge capacity of the composite cathode returns to 1012 mA h g^{-1} from 727 mA h g^{-1}, when the rate decreases from 0.5 C to 0.1 C. From the EIS of the composite electrode after 50 cycles (**Figure 26d**), there are wo depressed semicircles in the middle and high frequency region and an oblique line in low-frequency region. The semicircle from high to medium frequency represents the SEI layer resistance, which was adopted to protect the lithium anode and improve coulombic efficiency. The intermediate frequency semicircle represents charge transfer resistance between the electrolyte and cathode electrode, and the oblique line represents diffusion impedance [47].

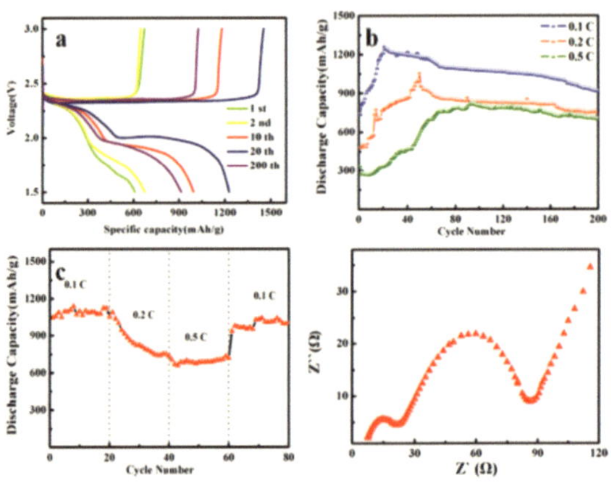

Figure 25. (a) Charge-discharge profile for the C-S@PANI composite at 0.1 C between 0.1 and 3.0 V (vs. Li/Li^+) at room temperature, (b) cycle performance of the C-S@PANI composite at 0.1, 0.2, and 0.5 C, respectively, (c) rate performance of C-S@PANI composite. (d) Electrochemical impedance spectra of the composite electrode after 50 cycles [47].

6.4. The charge-discharge mechanism of the C-S@PANI composite with PSN as cathode for high performance lithium-sulfur batteries

There are multiple reactions in the charge-discharge reaction of a lithium-sulfur battery. Sulphur can react with metallic lithium to form Li_2S with a large negative free energy change, which can be harnessed in a battery with a two-electron reaction [47]. As the redox reaction, sulfur entirely dissolves into the liquid electrolyte in the form of Li_2S_8 with S_8 ring as the initial form. The color of catholyte changed from red to green with decreasing chain length of polysulfides [52]. The final product is insoluble Li_2S_2 or Li_2S, which is deposited on the electrode surface, blocking further reaction. This process of inclusion complex disproportionation results in the "shuttle effect" [51]. The total reaction is as follows: $S_8 + 16Li \rightarrow 8Li_2S$. The PSN provides access to Li^+ ingress/egress for conversion of Li_2S to Li_2S_x (as illustrated in **Figure 26**).

The higher specific capacity and excellent cycle stability are achieved simultaneously, which can be attributed to the special microstructure of the C-S@PANI composite. At first, the C-S core-shell composites coated by the conductive PANI form a three-dimensional conducting network in **Figure 26a**. It is beneficial for enhancement of the rate capability of sulfur cathode, because the utilization of the active mass and dual conduction of Li^+ and electrons. Secondly, the much more mass of sulfur could be encapsulated into the PANI hollow spherical network in order to generate the S-C bonds act as the bridge between the PANI and sulfur or polysulfides, during heating treatment. The structure of hyperbranched, high aspect ratio, and multiple cross junctions can inhibit effectively the dissolution and migration of polysulfides into the electrolyte, thus effectively enhancing the cycling stability and the coulombic efficiency of the electrode at high current rates [47, 53]. In addition, the PANI hollow network could supply sufficient space to buffer the large volume expansion of sulfur during the charge-discharge process (**Figure 26b**). Thus, as the cathode material of C-S@PANI composite with PSN structure, the superior specific capacity, rate stability, and excellent cycling performance of the lithium sulfur battery should be improved.

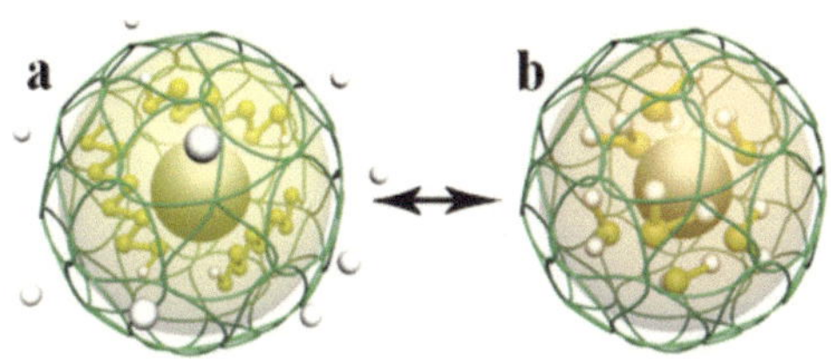

Figure 26. Li_2S (orange color) trapping within the PANI framework during lithiation/delithiation (yellow and gray are the sulfur and lithium) [47].

7. Conclusions

In this chapter, new kinds of cathode materials for lithium sulfur batteries have been designed and synthesized. Generally, sulfur cathodes have diversified through microstructure designing with various materials, including core-shell HPC spheres, graphene, GOs, polymers and their hybrids, which should effectively inhibit the dissolution and migration of polysulfides into the electrolyte, and volume expansion during discharge. In addition, as cathode material for Li-S secondary batteries, the as-prepared new kind cathodes which improve the cycling stability and the coulombic efficiency of the electrode at different current rates and temperature, exert outstanding electrochemical features. So, the future work in the electrolyte area for Li-S batteries should focus on inhibiting the dissolution of polysulfides in electrolytes while optimizing their ionic and charge conductivity without compromising properties during prolonged electrochemical cycles.

Acknowledgements

This work was supported by the State Scholarship Fund of China Scholarship Council (grant no. 201408220025), Science Foundation of Jilin Education Department (grant no. 2015118), and the Special Funds of Changchun University of Technology.

Author details

Lianfeng Duan[1*], Feifei Zhang[2] and Limin Wang[2]

*Address all correspondence to: duanlf@ccut.edu.cn

1 Key Laboratory of Advanced Structural Materials, Ministry of Education, Department of Materials Science and Engineering, Changchun University of Technology, Changchun, China

2 State Key Laboratory of Rare Earth Resource Utilization, Changchun Institute of Applied Chemistry, Chinese Academy of Sciences, Changchun, China

References

[1] A. Manthiram, Y. Fu, S. H. Chung, C. Zu, Y. S. Su. Rechargeable lithium-sulfur batteries. Chem Rev. 2014;114:11751–11787.

[2] G. Xu, B. Ding, J. Pan, P. Nie, L. Shen, X. Zhang. High performance lithium-sulfur batteries: advances and challenges. J Mater Chem A 2014;2:12662–12669.

[3] S. Zhang, K. Ueno, K. Dokko, M. Watanabe. Recent advances in electrolytes for lithium-sulfur batteries. Adv. Energy Mater. 2015;5:15–43.

[4] R. Wu, D. P. Wang, X. Rui, B. Liu, K. Zhou, A. W. Law, Q. Yan, J. Wei, Z. Chen. In-situ formation of hollow hybrids composed of cobalt sulfides embedded within porous

carbon polyhedra/carbon nanotubes for high-performance lithium-ion batteries. Adv Mater. 2015;27:3038–3044.

[5] M. A. Pope, I. A. Aksay. Structural design of cathodes for Li-S batteries. Adv. Energy Mater. 2015;5:1500124–1500146.

[6] R. Chen, T. Zhao, F. Wu. From a historic review to horizons beyond: lithium-sulphur batteries run on the wheels. Chem Commun. 2015;51:18–33.

[7] D. Herbert, J. Ulam. Electric dry cells and storage batteries. US3043896, 1962.

[8] J. R. Birk, R. K. Steunenberg. Chemical investigations of lithium-sulfur cells advances in chemistry. Am Chem Soc. 1975;140:186–202.

[9] H. Yamin, E. Peled. Electrochemistry of a nonaqueous lithium/sulfur cell. J Power Sources 1983;9:281–287.

[10] H. Yamin, J. Penciner, A. Gorenshtain, M. Elam, E. Peled. The electrochemical behavior of polysulfides in tetrahydrofuran. J Power Sources 1985;14:129–134.

[11] Y. F. A. Manthiram, Y. S. Su. Challenges and prospects of lithium-sulfur. Accounts Chem Res. 2012;46:1125–1134.

[12] R. Xu, J. Lu, K. Amine. Progress in mechanistic understanding and characterization techniques of Li-S batteries. Adv Energy Mater. 2015;5:1500430.

[13] D. W. Wang, Q. Zeng, G. Zhou, L. Yin, F. Li, H. M. Cheng, I. R. Gentle, G. Q. M. Lu. Carbon-sulfur composites for Li-S batteries: status and prospects. J Mater Chem A 2013;1:9382–9394.

[14] M. Barghamadi, A. S. Best, A. I. Bhatt, A. F. Hollenkamp, M. Musameh, R. J. Rees, T. Rüther. Lithium-sulfur batteries the solution is in the electrolyte, but is the electrolyte a solution? Energy Environ Sci. 2014;7:3902–3920.

[15] Y. Yang, G. Zheng, Y. Cui. Nanostructured sulfur cathodes. Chem Soc Rev. 2013;42:3018–3032.

[16] S. Chen, X. Huang, H. Liu, B. Sun, W. Yeoh, K. Li, J. Zhang, G. Wang. 3D Hyperbranched hollow carbon nanorod architectures for high-performance lithium-sulfur batteries. Adv Energy Mater. 2014;4:1301761–1301770.

[17] Y. Dong, S. Liu, Z. Wang, Y. Liu, Z. Zhao, J. Qiu. Sulfur-infiltrated graphene-backboned mesoporous carbon nanosheets with a conductive polymer coating for long-life lithium-sulfur batteries. Nanoscale 2015;7:7569–7573.

[18] H. S. Ryu, H. J. Ahna, K. W. Kim, J. H. Ahnb, J. Y. Lee. Discharge process of Li/PVdF/S cells at room temperature. J Power Sources 2006;153:360–364.

[19] H. Kim, H. D. Lim, J. Kim, K. Kang. Graphene for advanced Li/S and Li/air batteries. J Mater Chem A 2014;2:33–47.

[20] L. F. Nazar, M. Cuisinier, Q. Pang. Lithium-sulfur batteries. MRS Bull. 2014;39:436–442.

[21] Y. V. Mikhaylik, J. R. Akridge. Polysulfide shuttle study in the Li/S battery system. J Electrochem Soc. 2004;151:A1969–A1976.

[22] L. Suo, Y. S. Hu, H. Li, M. Armand, L. Chen. A new class of solvent-in-salt electrolyte for high-energy rechargeable metallic lithium batteries. Nat Commun. 2013;4:1481–1489.

[23] J. L. Wang, J. Yang, J. Y. Xie, N. X. Xu. A novel conductive polymer-sulfur composite cathode material for rechargeable lithium batteries. Adv Mater. 2002;14:963–965.

[24] B. Wang, Y. F. Wen, D. L. Ye, H. Yu, B. Sun, G. X. Wang, D. H. Jurcakova, L. Z. Wang. Dual protection of sulfur by carbon nanospheres and graphene sheets for lithium-sulfur batteries. Chem Eur J. 2014;20:5224–5230.

[25] H. L. Wang, Y. Yang, Y. Y. Liang, J. T. Robinson, Y. G. Li, A. Jackson, Y. Cui, H. J. Dai. Graphene-wrapped sulfur particles as a rechargeable lithium-sulfur battery cathode material with high capacity and cycling stability. Nano Lett. 2011;11:2644–2647.

[26] C. F. Zhang, H. B. Wu, C. Z. Yuan, Z. P. Guo, X. W. Lou. Confining sulfur in double-shelled hollow carbon spheres for lithium–sulfur batteries. Angew Chem. 2012;124:9730–9733.

[27] N. Jayaprakash, J. Shen, S. S. Moganty, A. Corona, L. A. Archer. Porous hollow carbon@sulfur composites for high-power lithium-sulfur batteries. Angew Chem. 2011;123:6026–6030.

[28] F. F. Zhang, G. Huang, X. X. Wang, Y. L. Qin, X. C. Du, D. M. Yin, F. Liang, L. M. Wang. Sulfur-impregnated core-shell hierarchical porous carbon for lithium-sulfur batteries. Chem Eur J. 2014;20:1–8.

[29] B. Zhang, X. Qin, G. R. Li, X. P. Gao. Enhancement of long stability of sulfur cathode by encapsulating sulfurinto micropores of carbon spheres. Energy Environ Sci. 2010;3:1531–1537.

[30] C. Tang, Q. Zhang, M. Q. Zhao, J. Q. Huang, X. B. Cheng, G. L. Tian, H. J. Peng, F. Wei. Nitrogen-doped aligned carbon nanotube/graphene sandwiches: facile catalytic growth on bifunctional natural catalysts and their applications as scaffolds for high-rate lithium-sulfur batteries. Adv Mater. 2014;26:6100–6105.

[31] G. He, S. Evers, X. Liang, M. Cuisinier, A. Garsuch, L. F. Nazar. Tailoring porosity in carbon nanospheres for lithium-sulfur battery cathodes. ACS Nano. 2013;7:10920–10930.

[32] G. M. Zhou, D. W. Wang, F. Li, P. X. Hou, L. C. Yin, C. Liu, G. Q. Lu, I. R. Gentle, H. M. Cheng. A flexible nanostructured sulphur-carbon nanotube cathode with high rate performance for Li-S batteries. Energy Environ Sci. 2012;5:8901–8906.

[33] J. C. Guo, Y. H. Xu, C. S. Wang. Sulfur-impregnated disordered carbon nanotubes cathode for lithium-sulfur batteries. Nano Lett. 2011;11:4288–4294.

[34] C. Lai, X. P. Gao, B. Zhang, T. Y. Yan, Z. Zhou. Synthesis and electrochemical performance of sulfur/highly porous carbon composites. J Phys Chem C 2009;113:4712–4716.

[35] J. Z. Wang, L. Lu, M. Choucair, J. A. Stride, X. Xu, H. K. Liu. Sulfur-graphene composite for rechargeable lithium batteries. J Power Sources 2011;196:7030–7034.

[36] D. C. Marcano, D. V. Kosynkin, J. M. Berlin, A. Sinitskii, Z. Sun, A. Slesarev, L. B. Alemany, W. Lu, J. M. Tour. Improved synthesis of graphene oxide. ACS Nano. 2010;4:4806–4814.

[37] F. F. Zhang, Y. H. Dong, Y. Huang, G. Huang, X. B. Zhang, L. M. Wang. Preparation and performance of a sulfur/graphene composite for rechargeable lithium-sulfur battery. J Phys Conf Ser. 2012;339:012003–012008.

[38] S. Evers, L. F. Nazar. Graphene-enveloped sulfur in a one pot reaction: a cathode with good coulombic efficiency and high practical sulfur content. Chem Commun. 2012;48:1233–1235.

[39] A. K. Geim, K. S. Novoselov. The rise of graphene. Nat Mater. 2007;6:183–191.

[40] L. Ji, M. Rao, H. I. Zheng, L. Zhang, Y. Li, W. Duan, J. Guo, E. J. Cairns, Y. Zhang. Graphene oxide as a sulfur immobilizer in high performance lithium/sulfur cells. J Am Chem Soc. 2011;133:18522–18525.

[41] X. Ji, K. Lee, L. Nazar. A highly ordered nanostructured carbon-sulphur cathode for lithium-sulphur batteries. Nat Mater. 2009;8:500–506.

[42] F. F. Zhang, X. B. Zhang, Y. H. Dong, L. M. Wang. Facile and effective synthesis of reduced graphene oxide encapsulated sulfur via oil/water system for high performance lithium sulfur cells. J Mater Chem. 2012;22:11452–11454.

[43] J. H. Shin, E. J. Cairns. N-Methyl-(n-butyl)pyrrolidinium bis(trifluoromethanesulfonyl)imide-LiTFSI-poly(ethylene glycol) dimethyl ether mixture as a Li/S cell electrolyte. J Power Sources 2008;177:537–545.

[44] J. Wang, Y. S. He, J. Yang. Sulfur-based composite cathode materials for high-energy rechargeable lithium batteries. Adv Mater. 2014;3:569–572.

[45] S. S. Zhang, J. A. Read. A new direction for the performance improvement of rechargeable lithium/sulfur batteries. J Power Sources 2012;200:77–82.

[46] J. X. Huang, S. Virji, B. H. Weiller, R. B. Kaner. Polyaniline nanofibers: facile synthesis and chemical sensors. J. Am. Chem. Soc. 2003;125:314–315.

[47] J. K. Wang, K. Q. Yue, X. D. Zhu, K. L. Wang, L. F. Duan. C-S@PANI composite with a polymer spherical network structure for high performance lithium-sulfur batteries. Phys Chem Chem Phys. 2016;18:261–266.

[48] W. Li, Q. Zhang, G. Zheng, Z. W. Seh, H. Yao, Y. Cui. Understanding the role of different conductive polymers in improving the nanostructured sulfur cathode performance. Nano Lett. 2013;13:5534–5540.

[49] L. F. Xiao, Y. L. Cao, J. Xiao, B. Schwenzer, M. H. Engelhard, L. V. Saraf, Z. M. Nie, G. J. Exarhos, J. Liu. A soft approach to encapsulate sulfur: polyaniline nanotubes for lithium-sulfur batteries with long cycle life. Adv Mater. 2012;24:1176–1181.

[50] W. Zhou, Y. Yu, H. Chen, F. J. DiSalvo, H. D. Abruña. Yolk-shell structure of polyaniline-coated sulfur for lithium-sulfur batteries. J Am Chem Soc. 2013; 135:16736–16743.

[51] C. Barchasz, F. Molton, C. Duboc, J. C. Leprêtre, S. Patoux, F. Alloin. Lithium/sulfur cell discharge mechanism: an original approach for intermediate species identification. Anal Chem. 2012;84:3973–3980.

[52] D. Marmorstein, T. H. Yu, K. A. Striebel, F. R. McLarnon, J. Hou, E. J. Cairns. Electrochemical performance of lithium/sulfur cells with three different polymer electrolytes. J Power Sources 2000;89:219–226.

[53] J. Song, T. Xu, M. L. Gordin, P. Y. Zhu, D. P. Lv, Y. B. Jiang,Y. S. Chen, Y. H. Duan, D. H. Wang. Nitrogen-doped mesoporous carbon promoted chemical adsorption of sulfur and fabrication of high-areal-capacity sulfur cathode with exceptional cycling stability for lithium-sulfur batteries. Adv Funct Mater. 2014;24:1243–1250.

Chapter 4

Metal Hydride-Based Materials as Negative Electrode for All-Solid-State Lithium-Ion Batteries

Liang Zeng, Koji Kawahito and Takayuki Ichikawa

Additional information is available at the end of the chapter

http://dx.doi.org/10.5772/62866

Abstract

The recently developed metal hydride (MH)-based material is considered to be a potential negative material for lithium-ion batteries, owing to its high theoretical Li storage capacity, relatively low volume expansion, and suitable working potential with very small polarization. However, it suffers from the slow kinetics, poor reversibility, and unfavourable cyclability in conventional organic liquid electrolyte systems, which enormously limit its practical application. In this chapter, we describe an all-solid-state battery system consisting of MH working electrode, $LiBH_4$ solid electrolyte, and Li metal counter electrode. The electrochemical properties of MgH_2 and TiH_2 composites are investigated, which showed much better performances using $LiBH_4$ as solid-state electrolyte than using conventional organic liquid electrolyte.

Keywords: Metal hydride, battery, solid state, conversion electrode, lithium-ion

1. Introduction

Currently, lithium-ion batteries (LIBs) with zero emission are considered to be the most important energy carrier in our daily life, which are absolutely imperative devices to power mobile phones, laptops, electric vehicles (EVs), and so on [1, 2]. Developing high-capacity materials for LIBs is necessary to satisfy the continuously rising need for energy. In the past tens of years, graphite is always used as negative electrode for commercial LIBs with an insufficient specific capacity (370 Ah kg^{-1}, 840 Ah L^{-1}) [3]. To overcome these restrictions, new concepts for the negative electrode must be developed. In other words, seeking suitable high-performance anode materials is essential and urgent.

In order to develop LIBs with high energy density, the conversion-type electrode materials had been intensively studied in recent years owing to their high Li storage capacity [2, 4]. In 2008, metal hydride (MH) was firstly reported as negative electrode for this purpose [5]. Compared to other conversion-type electrode materials, MH exhibits not only high theoretical Li storage capacities, but also relatively low volume expansions and suitable working potentials with very small polarization between charge and discharge [6]. The Li^+ insertion/extraction reactions for MHs can be written by the following hydride conversion reaction:

$$MH_x + xLi^+ + x\,e^- \leftrightarrow M + x\,LiH \quad (1)$$

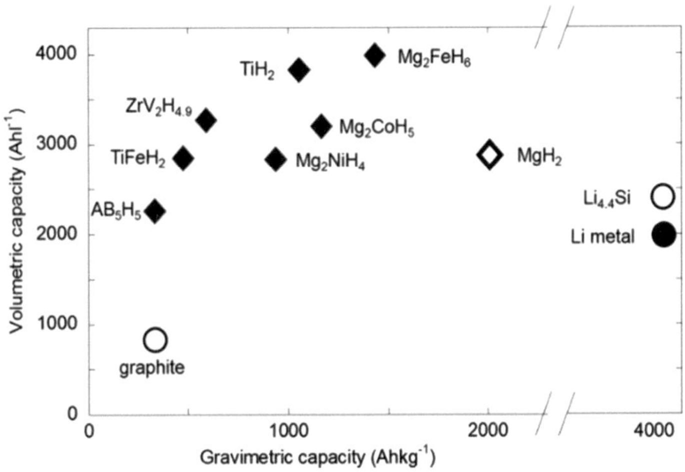

Figure 1. Theoretical gravimetric and volumetric capacities of metals and complex hydrides compared with those of graphite and other materials for negative electrodes. Reproduced from Ref. [6].

The theoretical gravimetric and volumetric capacities of some potential MHs as negative electrode for LIBs are shown in **Figure 1**. It can be seen that the capacities of all hydrides presented in this figure are larger than that of graphite. MgH_2 possesses the highest gravimetric capacity of 2038 Ah kg^{-1}, while Mg_2FeH_6 possesses the highest volumetric capacity of 3995 Ah L^{-1}. These large capacities lead MHs to a good candidate material for negative electrodes in LIBs for stationary as well as mobile applications for which the volumetric capacity plays a key role [6].

2. Metal hydride electrodes working in conventional organic liquid electrolyte systems

Among all of the MHs, MgH_2 is considered as the most promising candidate owing to its high theoretical Li storage capacity of 2038 mAh g^{-1}, a suitable working potential of around 0.5 V

http://dx.doi.org/10.5772/62866

(vs Li^+/Li), a relatively small volume change of 83%, and so on [5, 7]. The Li^+ insertion/extraction reaction of MgH_2 can be written by the following hydride conversion:

$$MgH_2 + 2Li^+ + 2e^- \leftrightarrow Mg + 2LiH \quad (2)$$

Further lithiation results in Li–Mg alloying reaction, which is partially irreversible.

$$Mg + xLi^+ + xe^- \leftrightarrow MgLi_x \quad (3)$$

For instance, the MgH_2 electrode working in 1 M $LiPF_6$ (DMC:EC=1:1) showed a reversible capacity of 1480 mAh g^{-1} in the first cycle; however, the capacity fades rapidly and reduces to less than 200 mAh g^{-1} after only 10 cycles. Sustained capacities could be achieved only by using a special copper foam current collector for particle confinement and limiting the reaction to 520 mAh g^{-1} [5]. A more recent report shows that a reversible capacity of about 500 mAh g^{-1} can be obtained after 20 discharge–charge cycles at a rate of 0.05 C with using nanoconfinement technique for MgH_2, but the coulombic efficiency of the first cycle is less than 50% [8].

After the discovery of MH as negative material for LIBs by Oumellal et al, some other groups followed their work and extended the research to many other hydrides, that is, TiH_2, TiNiH, Mg_2FeH_6, AlH_3, $LiAlH_4$, $NaAlH_4$, and so on [6, 8–24]. The electrochemical performances of these materials are summarized in **Table 1**.

Sample	Theoretical capacity (mAh g^{-1})	Reversible capacity of 1st cycle (mAh g^{-1})	Current density (C)	Cycling performance	Composition of electrode	Voltage range (V vs Li/Li^+)	Reference
MgH_2	2038	1480	1/20	520 mAh g^{-1} after 50 cycles*1	MgH_2:Carbon SP = 85:15	0.15–3.0	[5]
MgH_2	2038	1000	1/20	Less than 50 mAh g^{-1} after 10 cycles	MgH_2:PVdF:Super P carbon = 7:1:2	0.2–2.5	[9]
MgH_2	2038	1900	1/20	542 mAh g^{-1} after 40 cycles	MgH_2:CMC-f:$C_{7,460}$ carbon = 1:1:1	0.005–3.0	[10]
TiH_2	1074	1072*2	1/200	N/A	TiH_2:$C_{150,480}$ carbon = 10:1	0.005–3.0	[11]
TiNiH	248	251*3	1/10	N/A	TiNiH:Super P carbon = 10:1	0.005–3.0	[12]
Mg_2FeH_6	1454	1577*4	1/60	N/A	Mg_2FeH_6:$C_{150,480}$ carbon = 10:1	0.005–3.0	[13]
Mg_2CoH_5	1189	1300*5	1/50	N/A	Mg_2CoH_5:$C_{150,480}$ carbon = 10:1	0.005–3.0	[13]

Sample	Theoretical capacity (mAh g^{-1})	Reversible capacity of 1st cycle (mAh g^{-1})	Current density (C)	Cycling performance	Composition of electrode	Voltage range (V vs Li/Li^+)	Reference
Mg_2NiH_4	962	866*6	1/40	N/A	Mg_2NiH_4:$C_{150,480}$ carbon = 10:1	0.005–3.0	[13]
MgH_2	2038	950	1/20	500 mAh g^{-1} after 20 cycles	50MgH_2@HSAG-500:SP carbon = 80:20	0.005–3.0	[8]
$LiAlH_4$	2119	189	1/20	N/A	$LiAlH_4$:PVdF:Super P carbon = 5:2:3	0.01–2.5	[14]
Li_3AlH_6	1493	198	1/20	N/A	Li_3AlH_6:PVdF:Super P carbon = 5:2:3	0.01–2.5	[14]
TiH_2/C	1074	160	1/100	N/A	TiH_2:PVdF:Super P carbon = 5:2:3	0.01–2.0	[15]
$Mg_{0.7}Ti_{0.3}H_2$	1748	1540	1/20	530 mAh g^{-1} after 7 cycles	MgH_2:TiH_2:$C_{t,x}$ = 7:3:1	0.005–3.0	[16]
AlH_3	2678	1100	1/30	393 mAh g^{-1} after 10cycles	AlH_3:AB:CMC-f = 1:1:1	0.005–3.0	[17]
$NaAlH_4$	1984	770	1/30	250 mAh g^{-1} after 10 cycles	$NaAlH_4$:AB:CMC-f = 1:1:1	0.005–3.0	[17]

*1 Cycled using a pressed copper foam current collector.
*2*3*4*5*6 Only discharge capacity was shown.

Table 1. Electrochemical performance of metal hydrides in conventional organic liquid electrolyte (1M $LiPF_6$ (DMC:EC=1:1).

In spite of numerous advantages of MH mentioned above, the research on MH as negative electrode materials did not get enough attention since it suffers from the slow kinetics, poor reversibility, and unfavourable cyclability in conventional organic liquid electrolyte systems, which enormously limits its practical application.

3. MgH_2 composite electrode working with $LiBH_4$ solid-state electrolyte

From the above results, it can be inferred that the conventional organic liquid electrolyte systems might not be suitable for MH electrodes, thus an idea of using solid-state Li-ion conductor as electrolyte had come out.

There are many types of solid-state Li-ion conductors, including LISICON-type, Garnet-type, Perovskite-type, NASICON-type lithium-ion conductors, and glassy and glass-ceramic systems made of oxides and sulphides [25, 26]. When seeking a suitable medium to achieve a reversible conversion reaction by MH shown by Eq. (2), the Li^+ conductivity is one side to be considered, and another important factor is the H^- conductivity. As we know, the MgH_2–

$LiBH_4$ composite is a well-known material in hydrogen storage research field. In this system, the mobility of H^- in MgH_2 can be strongly enhanced by $LiBH_4$, due to a superior hydrogen exchange effect between MgH_2 and $LiBH_4$ [27]. That is to say, this effect may lead to a significant improvement of H^- conductivity when using this material as negative electrode for LIB. In the meantime, it was reported that $LiBH_4$ showed remarkable high Li-ion conductivity to the order of 10^{-3} S cm^{-1} at its HT phase (>115°C) with a very stable operating window between 0 and 5 V vs Li^+/Li, which can be used as solid-state electrolyte for LIBs [26, 28–30]. Considering the above two effects, when using MgH_2–$LiBH_4$ as negative electrode and $LiBH_4$ as solid-state electrolyte, the hydride conversion reaction of Eq. (2) could be dramatically accelerated, which may lead to much better battery performance for reversibility and cyclability.

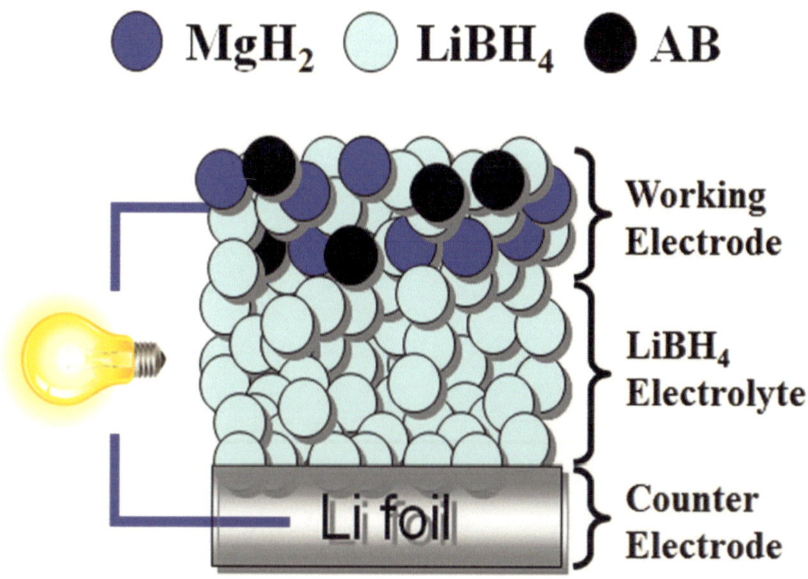

Figure 2. The schematic illustration of the MgH_2–$LiBH_4$ | $LiBH_4$ | Li all-solid-state battery.

In this part, a battery cell constructed of MgH_2–$LiBH_4$ | $LiBH_4$ | Li (illustrated in **Figure 2**) had been fabricated for electrochemical tests. 1 mol% Nb_2O_5-doped MgH_2 (named c-MgH_2 hereafter) was also used in this study for comparison, which had been proved to have positive effect in battery performance [20, 21, 23]. The working electrode was prepared by simply ball-milling MgH_2/c-MgH_2, $LiBH_4$, and conductive carbon (acetylene black) with no further optimization. The thermal stability of MgH_2–$LiBH_4$ composite was measured by thermogravimetry with differential thermal analysis (TG–DTA), which shows a weight loss of less than 1 wt% in the temperature of RT–275°C (**Figure 3**). To obtain the fast Li^+ conduction from $LiBH_4$, all electrochemical measurements in this study were conducted at 120°C. The electrode evolution upon discharge–charge process at different stages was investigated by means of powder X-ray diffraction (XRD).

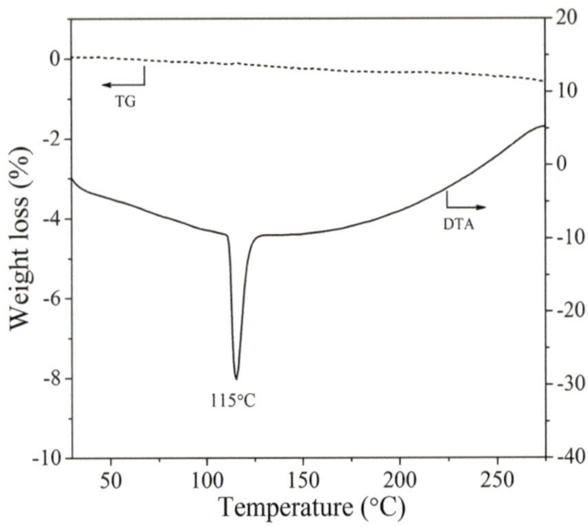

Figure 3. The TG–DTA result of the MgH_2–$LiBH_4$ composite heating up to 275°C.

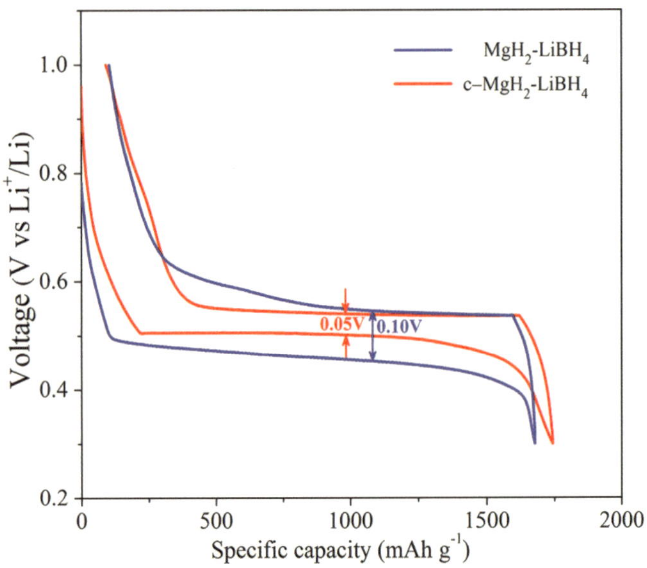

Figure 4. The first galvanostatic discharge–charge curves for MgH_2–$LiBH_4$ and c-MgH_2–$LiBH_4$ composite electrodes in the voltage range of 0.3–1.0 V at a current density of 100 mA g^{-1} at 120°C, respectively.

Figure 4 shows the first galvanostatic discharge–charge curves for MgH_2–$LiBH_4$ and c-MgH_2–$LiBH_4$ composite electrodes using $LiBH_4$ as solid-state electrolyte in the voltage range of 0.3–1.0 V vs Li^+/Li at a current density of 100 mA g^{-1}. It can be seen that on the discharge curve (Li^+ incorporation into MgH_2), the potential gradually dropped to near 0.5 V, followed by showing a long flat plateau for c-MgH_2 electrode and a slope for MgH_2 electrode, corresponding to the hydride conversion reaction shown by Eq. (2). The discharge process was cut at 0.3 V to avoid the Li–Mg alloying reaction: $Mg + x\ Li^+ + x\ e^- \leftrightarrow MgLi_x$, which is a partial irreversible reaction and would weaken the cyclability of MgH_2. On the charge curve (Li^+ extraction), a plateau slightly above 0.5 V was observed for both samples. The polarization between Li insertion and extraction was 0.1 V for MgH_2 electrode and 0.05 V for c-MgH_2 electrode, which are much smaller than that of the MgH_2 electrodes working with organic liquid-based electrolyte in some previous reports [5, 9]. Such a small polarization could be owing to the high working temperature of 120°C, which brings better kinetic properties to the reaction. The initial reversible capacity and coulombic efficiency were 1650 mAh g^{-1} and 94.7% for c-MgH_2 electrode and 1572 mAh g^{-1} and 93.7% for MgH_2 electrode. The above results indicated that the existence of $LiBH_4$ significantly promoted the hydride conversion reaction due to its dual effects on Li^+ conductivity and H^- conductivity. In addition, the Nb_2O_5 doping to MgH_2 further improves the reaction kinetics, resulting in better reversibility and higher initial coulombic efficiency.

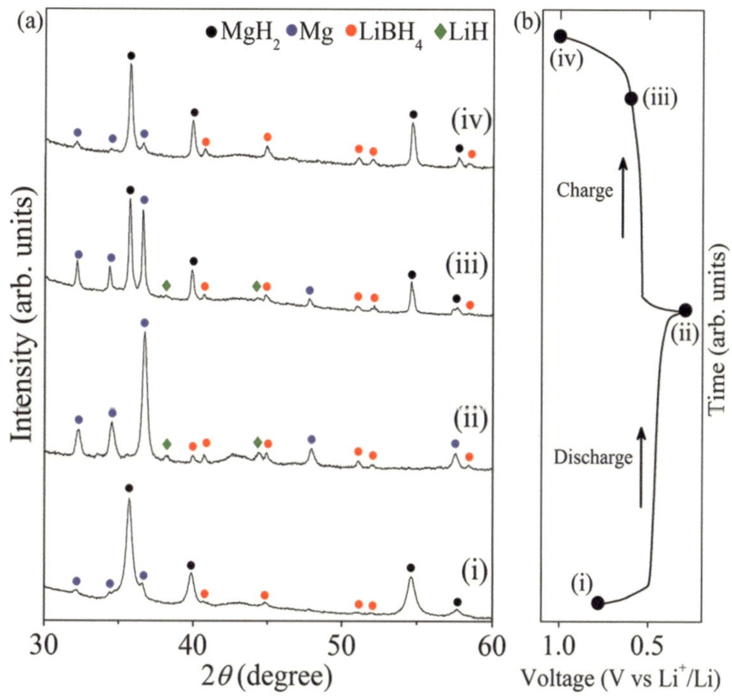

Figure 5. Ex situ XRD patterns of MgH_2–$LiBH_4$ composite electrode evolution upon the first electrochemical discharge–charge process at different stages.

The ex situ XRD measurement had been carried out at different stages upon the initial discharge–charge process shown by **Figure 5**. The $LiBH_4$ phase appeared in all the stages since it was included in the working electrode. At open circuit voltage stage (i), a small amount of Mg phase was detected in the XRD pattern as an impurity, which came from the as-received commercial MgH_2 sample. As Li reacted with MgH_2, the LiH and Mg peaks appeared while MgH_2 peaks disappeared at 0.3 V (ii), indicating the hydride conversion reaction shown by

Eq. (2) had occurred. When recharging the same battery cell, clear MgH_2 peaks reappeared with weakened LiH and Mg peaks at the intermediate stage (iii). Finally, the LiH phase totally disappeared with a very small amount of Mg at 1.0 V (iv), which is equivalent to stage (i). The above results verified the good reversibility of the hydride conversion reaction. Interestingly, a new phase of $Mg(BH_4)_2$ was found (**Figure 6**) when further charging the battery cell to 2.0 V, which implies a $Mg(BH_4)_2$ generation reaction might have occurred between 1.0 and 2.0 V shown by Eq. (4).

$$Mg + 2\,LiBH_4 \rightarrow Mg(BH_4)_2 + 2\,Li^+ + 2\,e^- \tag{4}$$

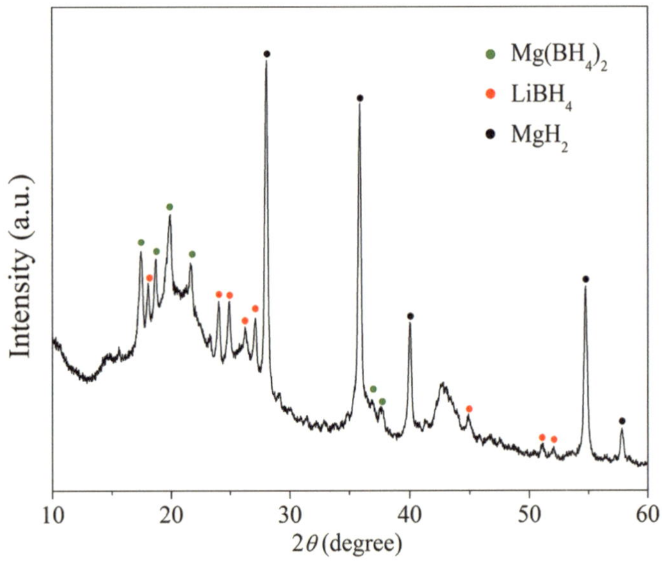

Figure 6. The XRD result of MgH_2–$LiBH_4$ composite electrode charged to 2.0 V.

To avoid this undesirable reaction, the galvanostatic discharge–charge cycling tests were performed between 0.3 and 1.0 V at a current density of 800 mA g^{-1} as shown in **Figure 7**. It shows that the MgH_2–$LiBH_4$ electrode delivered a highly reversible capacity of 1488 mAh g^{-1} in the first cycle; however, only 270 mAh g^{-1} remained after 93 cycles. In the meantime, a reversible capacity of 1586 mAh g^{-1} is shown in the first cycle for the Nb_2O_5-doped electrode, which retained at 700 mAh g^{-1} after cycling for 100 times. This result suggests the cyclability of the MgH_2–$LiBH_4$ electrode is also improved by the Nb_2O_5-doping effect. The galvanostatic discharge–charge curves of Nb_2O_5-doped electrode are shown by **Figure 8**. It can be observed that the plateau voltage of the discharge and charge profiles almost has no changes during 100 cycles with coulombic efficiency over 99.5%, which indicated that the working electrode operated properly between 0.3 and 1.0 V, namely, there was no side reaction during the discharge–charge cycles. Moreover, the suitable Li insertion/extraction voltage of around 0.5 V (vs Li^+/Li) and the small polarization are particularly remarkable, making it a promising candidate as negative electrode material for LIBs.

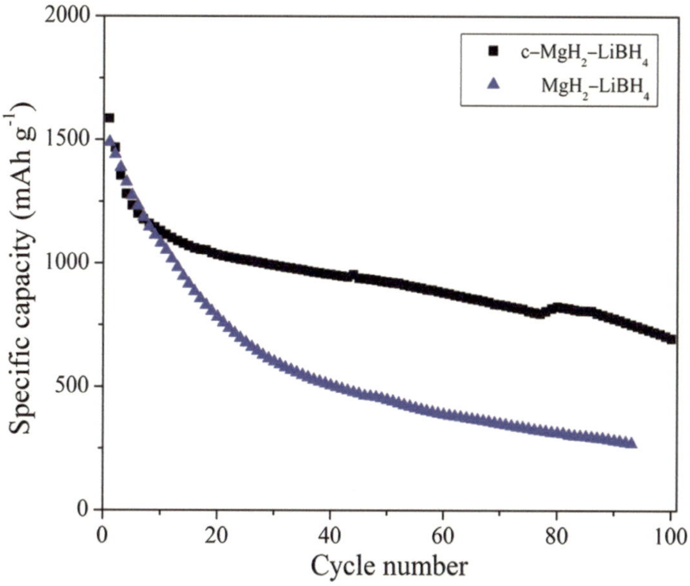

Figure 7. Cycling performances of MgH_2–$LiBH_4$ and c-MgH_2–$LiBH_4$ composite electrodes in the voltage range of 0.3–1.0 V at a current density of 800 mA g^{-1} at 120°C.

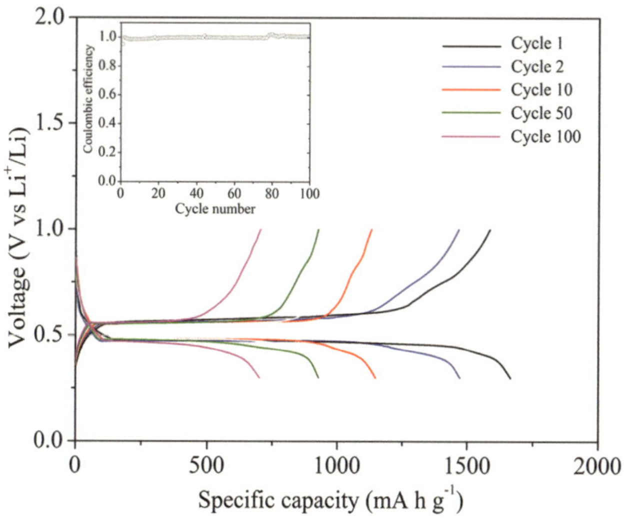

Figure 8. Galvanostatic discharge–charge curves for c-MgH_2–$LiBH_4$ composite electrode in the voltage range of 0.3–1.0 V at a current density of 800 mA g^{-1} at 120°C. Inset shows the coulombic efficiency for 100 cycles.

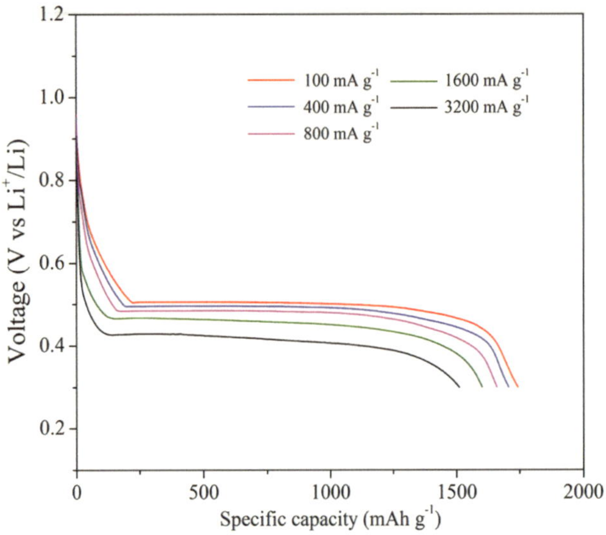

Figure 9. Rate performance of c-MgH_2–$LiBH_4$ composite electrodes at the current densities of 100–3200 mA g^{-1} at 120°C, respectively.

The rate capability of the Nb_2O_5-doped electrode (c-MgH_2–$LiBH_4$) was also investigated as shown by **Figure 9**. It can be seen that the discharge capacity was 1510, 1600, 1659, 1704, and 1742 mAh g^{-1} at the current density of 3200, 1600, 800, 400, and 100 mA g^{-1}, respectively, indicating that this sample possessed excellent rate capability. The fast kinetics of this material could be ascribed to rapid Li^+ and H^- diffusion in the working electrode and on the interface between electrode and electrolyte. It is worth noting that the plateau voltage gradually decreased along with increase in the current density, which was caused by the insufficiency of electrical conductivity at high discharging rate. As we know, the electrical conductivity of electrode material can be remarkably improved via surface modification technique, for example, carbon coating [31], which can be also applied in this system for future working direction in order to obtain better high rate performance.

4. TiH_2 electrode working with $LiBH_4$ solid-state electrolyte

Another promising candidate of MHs as negative electrode is titanium hydride (TiH_2), which has a theoretical capacity of 1074 mAh g^{-1} with a low working potential of 0.163 V vs Li^+/Li by the following reaction:

$$TiH_2 + 2\,Li^+ + 2\,e^- \leftrightarrow Ti + 2\,LiH \quad (5)$$

In the viewpoint of making high-power battery, TiH_2 is preferable to MgH_2 as negative electrode since its theoretical working potential is much lower than MgH_2, which leads to a higher voltage for a full battery cell. The electrochemical Li insertion/extraction mechanism of TiH_2 was reported by Oumellal et al. [11]. However, the cyclic properties have not been reported yet due to the poor reversibility in conventional organic liquid-based electrolyte.

In this part, we had prepared the TiH_2–$LiBH_4$ composite as the negative electrode together with $LiBH_4$ as the solid-state electrolyte for an all-solid-state battery cell, in order to obtain good cyclic performance for TiH_2. This system is similar to the above content for MgH_2 electrode, which shows favourable electrochemical performances. The phase evolution of TiH_2 electrode upon charge–discharge in the solid-state cell, cyclic properties, and rate performance are investigated.

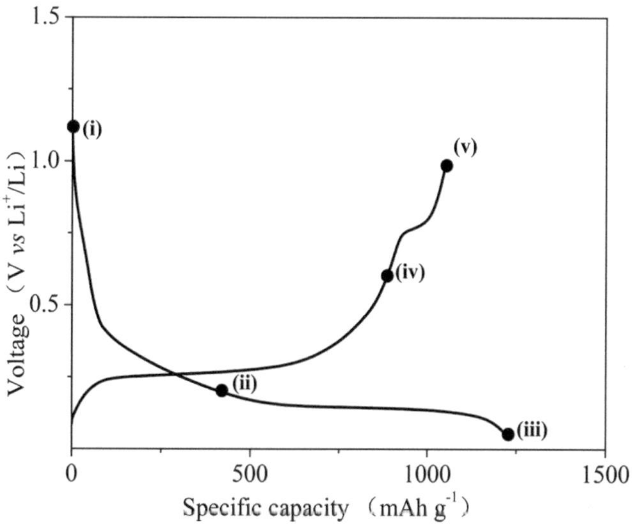

Figure 10. Initial galvanostatic charge–discharge curve for TiH_2–$LiBH_4$ electrode in the voltage range of 2.0–0.05 V at a current density of 400 mAh g^{-1} at 120°C.

Figure 10 shows the initial charge–discharge curves of the TiH_2 electrode between 1.0 and 0.05 V using $LiBH_4$ solid electrolyte under the current density of 400 mA g^{-1} at 120°C. The Li insertion curve shows slope and plateau regions: slope region (0.4–0.15 V) and plateau region (0.15 V). From some previously reported results [11], it was considered that the slope region indicated fcc-TiH_2 reacted with lithium to form LiH and partially transformed into a distorted orthorhombic phase fco-TiH_{2-x}, and slope region indicated that both of fcc-TiH_2 and fco-TiH_{2-x} react with lithium to form Ti and LiH ($TiH_2 + 2Li^+ + 2e^- \rightarrow Ti + 2LiH$). Unlike the case of MgH_2, no plateau could be observed below 0.15V, indicating that Ti did not electrochemically react with Li to form Li–Ti alloy. During the initial Li insertion, the capacity of 1225 mAh g^{-1} was obtained, while the theoretical capacity was 1074 mAh g^{-1}. This overcapacity could be due to the existence of acetylene black in the electrode, which contributes on the Li storage capacity as an anode.

On the Li extraction curve, the plateau corresponding to de-lithiation process of TiH_2 conversion reaction was observed at higher voltages than 0.15 V, indicating the higher polarization of electrode compared with the case of MgH_2 system [23]. Furthermore, a short plateau at 0.75 V was observed and the whole Li extraction curve showed 1052 mAh g^{-1} corresponding to 86% coulombic efficiency.

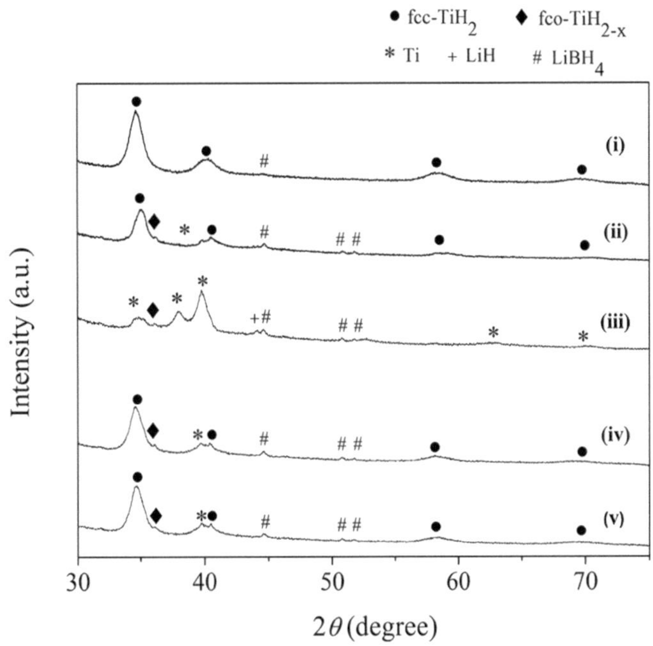

Figure 11. Ex situ XRD patterns of TiH_2–$LiBH_4$ at the various states of charge.

In order to identify the electrochemical reaction, ex situ XRD measurements were performed at the different states of charges as shown in **Figure 11**. It is noteworthy that all the profiles shown in **Figure 11** contain peaks corresponding to $LiBH_4$, which should always be included in the electrodes. At the initial state (i), the broad peaks corresponding to TiH_2 were observed due to the ball-milling effect. After the Li insertion (ii), the broad peaks slightly shifted to higher angle, indicating that the TiH_2 structure should be slightly shrunk due to partial extraction of hydrogen. And a small peak corresponding to fco-TiH_{2-x} and Ti were observed. With further lithiation (iii), it was clearly observed that the TiH_2 changed to Ti and LiH. For the Li extraction process, H atoms were transferred from LiH to Ti. Then, as shown in **Figure 10** (iv, v), peaks corresponding to TiH_2 reappeared, then the Ti peak was strongly reduced. These facts indicate that TiH_2 electrode reveals high electrochemical performance by using the $LiBH_4$ solid-state electrolyte. The combination of hydride and borohydride for anode and electrolyte, respectively, could lead to a better performance.

Unfortunately, the reaction at 0.75 V as shown in **Figure 10** was not clarified, yet even though we have performed the XRD measurement at various states of charge. As one of the possible reasons, it could be caused by the Li extraction process from acetylene black, which cannot be detected by XRD due to its amorphous feature.

http://dx.doi.org/10.5772/62866

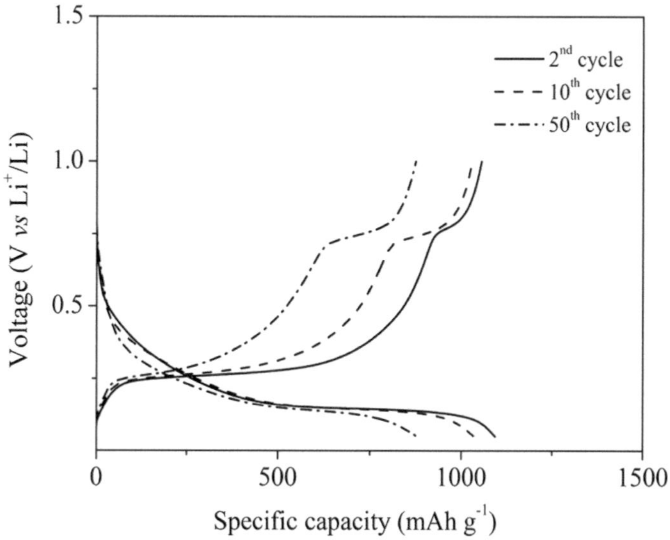

Figure 12. Cyclic performance of the TiH_2–$LiBH_4$ composite electrode in the voltage range of 1.0–0.05 V at a current density of 400 mA g^{-1} at 120°C.

Figure 12 shows the cyclic performance of TiH_2–$LiBH_4$ composite electrode in the voltage range of 1.0–0.05 V at a current density of 400 mA g^{-1} at 120°C. Almost the same profile as the initial cycle was observed in later cycles. The Li insertion capacities of the 2nd, 10th, and 50th cycles were 1094, 1035, and 878 mAh g^{-1}, respectively, which corresponding to a capacity retention of about 80% for 50 charge–discharge cycles. This result revealed for the first time that the

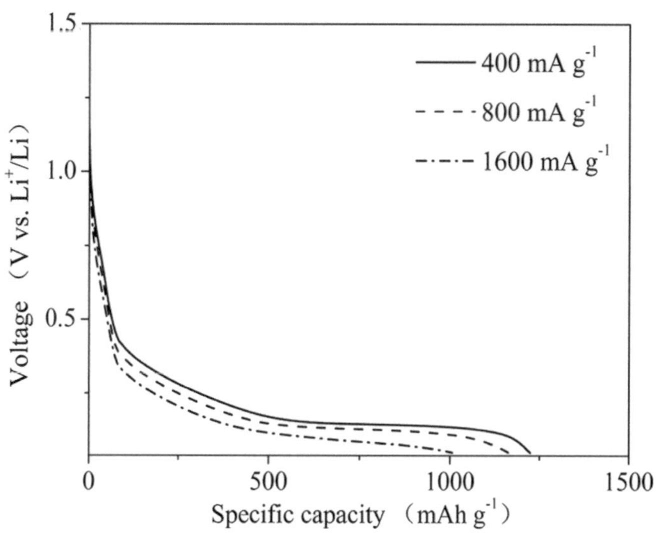

Figure 13. Rate capability of the TiH_2–$LiBH_4$ composite electrode at a current density of 400, 800, and 1600 mA g^{-1} at 120°C.

charge–discharge process of TiH_2 electrode is recyclable, which makes it a promising candidate for LIBs.

The rate capability of the TiH_2 composite electrode was also investigated as shown in **Figure 13**. All the discharge profiles in this figure were created by the initial discharge cycle of independent cells at different current densities of 400, 800, and 1600 mA g^{-1}, which are corresponding to about 1/3 C, 2/3 C, and 4/3 C, respectively. It can be seen that discharge capacity was 1225, 1165, and 1007 mAh g^{-1} at the current density of 400, 800, and 1600 mA g^{-1}, respectively, indicating a very good rate capability of TiH_2 electrode in this system. For comparison, the charge–discharge for TiH_2 in a conventional organic electrolyte could be performed only at 10 mA g^{-1} [6]. Like in the MgH_2 system, the fast kinetics of TiH_2 could also be ascribed to rapid Li^+ and H^- diffusion in the electrode material and on the electrode–electrolyte interface at a high working temperature of 120°C.

5. Summary

In summary, we have demonstrated an all-solid-state battery system consisting of MH working electrode, $LiBH_4$ solid electrolyte, and Li metal counter electrode. The MgH_2–$LiBH_4$ and TiH_2–$LiBH_4$ composite electrodes showed much better performances using $LiBH_4$ as solid-state electrolyte than using conventional organic liquid electrolyte in some previous reports. $LiBH_4$ here acts as multifunctional ion conductor, promotes not only Li^+ but also H^- conductivity, resulting in much higher reversibility for the hydride conversion reaction. In addition, Nb_2O_5 doping shows significant improvement on cyclability of MgH_2. The suitable operating voltage, large reversible capacity, and the very small polarization are particularly remarkable to make MH a promising negative electrode material for all-solid-state LIBs. Besides, not only $LiBH_4$ but also a large number of hydride-based solid-state alternatives for example, $Li_m(BH_4)_nX$ (X = Cl, Br, I) [29, 32], $LiBH_4$–$LiNH_2$ [29, 33], and $LiAlH_4$/Li_3AlH_6 [34] can also be considered for this system. The working temperature can be altered while using these various substitutions, some of which even show high Li-ion conductivity at room temperature. This work shows a new direction to search for potential high-performance negative electrode materials for LIBs. The electrochemical performance of numerous MH-based materials will be investigated in a similar system as in this article in the future.

Author details

Liang Zeng[1], Koji Kawahito[2] and Takayuki Ichikawa[1,3*]

*Address all correspondence to: tichi@hiroshima-u.ac.jp

1 Institute for Advanced Materials Research, Hiroshima University, Japan

2 Graduate School of Advanced Sciences of Matter, Hiroshima University, Japan

3 Graduate School of Integrated Arts and Sciences, Hiroshima University, Japan

References

[1] Armand M, Tarascon JM. Building better batteries. Nature. 2008;451:652–657. DOI: 10.1038/451652a

[2] Bruce PG, Scrosati B, Tarascon JM. Nanomaterials for rechargeable lithium batteries. Angew. Chem. Int. Ed. 2008;47:2930–2946. DOI: 10.1002/anie.200702505

[3] Arico AS, Bruce P, Scrosati B, Tarascon JM, Van Schalkwijk W. Nanostructured materials for advanced energy conversion and storage devices. Nat. Mater. 2005;4:366–377. DOI: 10.1038/Nmat1368

[4] Reddy MV, Rao GVS, Chowdari BVR. Metal Oxides and Oxysalts as Anode Materials for Li Ion Batteries. Chem. Rev. 2013;113:5364–5457. DOI: 10.1021/Cr3001884

[5] Oumellal Y, Rougier A, Nazri GA, Tarascon JM, Aymard L. Metal hydrides for lithium-ion batteries. Nat. Mater. 2008;7:916–921. DOI: 10.1038/Nmat2288

[6] Aymard L, Oumellal Y, Bonnet JP. Metal hydrides: an innovative and challenging conversion reaction anode for lithium-ion batteries. Beilstein J. Nanotech. 2015;6:1821–1839. DOI: 10.3762/bjnano.6.186

[7] Oumellal Y, Rougier A, Tarascon JM, Aymard L. 2LiH+M (M = Mg, Ti): New concept of negative electrode for rechargeable lithium-ion batteries. J. Power Sources. 2009;192:698–702. DOI: 10.1016/j.jpowsour.2009.03.003

[8] Oumellal Y, Zlotea C, Bastide S, Cachet-Vivier C, Leonel E, Sengmany S, Leroy E, Aymard L, Bonnet JP, Latroche M. Bottom-up preparation of MgH2 nanoparticles with enhanced cycle life stability during electrochemical conversion in Li-ion batteries. Nanoscale. 2014;6:14459–14466. DOI: 10.1039/C4nr03444a

[9] Brutti S, Mulas G, Piciollo E, Panero S, Reale P. Magnesium hydride as a high capacity negative electrode for lithium ion batteries. J. Mater. Chem. 2012;22:14531–14537. DOI: 10.1039/C2jm31827j

[10] Zaidi W, Oumellal Y, Bonnet JP, Zhang J, Cuevas F, Latroche M, Bobet JL, Aymard L. Carboxymethylcellulose and carboxymethylcellulose-formate as binders in MgH2-carbon composites negative electrode for lithium-ion batteries. J. Power Sources. 2011;196:2854–2857. DOI: 10.1016/j.jpowsour.2010.11.048

[11] Oumellal Y, Zaidi W, Bonnet JP, Cuevas F, Latroche M, Zhang J, Bobet JL, Rougier A, Aymard L. Reactivity of TiH2 hydride with lithium ion: Evidence for a new conversion mechanism. Int. J. Hydrogen Energy. 2012;37:7831–7835. DOI: 10.1016/j.ijhydene.2012.01.107

[12] Bououdina M, Oumellal Y, Dupont L, Aymard L, Al-Gharni H, Al-Hajry A, Maark TA, De Sarkar A, Ahuja R, Deshpande MD, Qian Z, Rahane AB. Lithium storage in amorphous TiNi hydride: Electrode for rechargeable lithium-ion batteries. Mater. Chem. Phys. 2013;141:348–354. DOI: 10.1016/j.matchemphys.2013.05.021

[13] Zaidi W, Bonnet JP, Zhang J, Cuevas F, Latroche M, Couillaud S, Bobet JL, Sougrati MT, Jumas JC, Aymard L. Reactivity of complex hydrides Mg2FeH6, Mg2CoH5 and Mg2NiH4 with lithium ion: Far from equilibrium electrochemically driven conversion reactions. Int. J. Hydrogen Energy. 2013;38:4798–4808. DOI: 10.1016/j.ijhydene.2013.01.157

[14] Silvestri L, Forgia S, Farina L, Meggiolaro D, Panero S, La Barbera A, Brutti S, Reale P. Lithium Alanates as Negative Electrodes in Lithium-Ion Batteries. Chemelectrochem. 2015;2:877–886. DOI: 10.1002/celc.201402440

[15] Vitucci F, Paolone A, Brutti S, Munaò D, Silvestri L, Panero S, Reale P. H2 thermal desorption and hydride conversion reactions in Li cells of TiH2/C amorphous nanocomposites. J. Alloys Compd. 2015;645:S46–S50. DOI: 10.1016/j.jallcom.2015.01.232

[16] Huang L, Aymard L, Bonnet JP. MgH2-TiH2 mixture as an anode for lithium-ion batteries: synergic enhancement of the conversion electrode electrochemical performance. J. Mater. Chem. A. 2015;3:15091–15096. DOI: 10.1039/c5ta02545a

[17] Teprovich JA, Zhang JX, Colon-Mercado H, Cuevas F, Peters B, Greenway S, Zidan R, Latroche M. Li-Driven Electrochemical Conversion Reaction of AIH3, LiAIlH4, and NaAlH4. J. Phys. Chem. C. 2015;119:4666–4674. DOI: 10.1021/jp5129595

[18] Zhang JX, Zaidi W, Paul-Boncour V, Provost K, Michalowicz A, Cuevas F, Latroche M, Belin S, Bonnet JP, Aymard L. XAS investigations on nanocrystalline Mg2FeH6 used as a negative electrode of Li-ion batteries. J. Mater. Chem. A. 2013;1:4706–4717. DOI: 10.1039/C3ta01482g

[19] Meggiolaro D, Gigli G, Paolone A, Vitucci F, Brutti S. Incorporation of Lithium by MgH2: An Ab Initio Study. J. Phys. Chem. C. 2013;117:22467–22477. DOI: 10.1021/jp404993z

[20] Ikeda S, Ichikawa T, Kawahito K, Hirabayashi K, Miyaoka H, Kojima Y. Anode properties of magnesium hydride catalyzed with niobium oxide for an all solid-state lithium-ion battery. Chem. Commun. 2013;49:7174–7176. DOI: 10.1039/C3cc43987a

[21] Ikeda S, Ichikawa T, Goshome K, Yamaguchi S, Miyaoka H, Kojima Y. Anode properties of Al2O3-added MgH2 for all-solid-state lithium-ion batteries. J. Solid State Electr. 2015;19:3639–3644. DOI: 10.1007/s10008-015-2959-8

[22] Meggiolaro D, Gigli G, Paolone A, Reale P, Doublet ML, Brutti S. Origin of the Voltage Hysteresis of MgH2 Electrodes in Lithium Batteries. J. Phys. Chem. C. 2015;119:17044–17052. DOI: 10.1021/acs.jpcc.5b04615

[23] Zeng L, Kawahito K, Ikeda S, Ichikawa T, Miyaoka H, Kojima Y. Metal hydride-based materials towards high performance negative electrodes for all-solid-state lithium-ion batteries. Chem. Commun. 2015;51:9773–9776. DOI: 10.1039/c5cc02614h

[24] Kawahito K, Zeng L, Ichikawa T, Miyaoka H, Kojima Y. Electrochemical performance of titanium hydride for bulk-type all-solid-state lithium-ion batteries. Mater. Trans. DOI: 10.2320/matertrans.M2016024

[25] Jung YS, Oh DY, Nam YJ, Park KH. Issues and Challenges for Bulk-Type All-Solid-State Rechargeable Lithium Batteries using Sulfide Solid Electrolytes. Isr. J. Chem. 2015;55:472–485. DOI: 10.1002/ijch.201400112

[26] Takahashi K, Hattori K, Yamazaki T, Takada K, Matsuo M, Orimo S, Maekawa H, Takamura H. All-solid-state lithium battery with LiBH4 solid electrolyte. J. Power Sources. 2013;226:61–64. DOI: 10.1016/j.jpowsour.2012.10.079

[27] Zeng L, Miyaoka H, Ichikawa T, Kojima Y. Superior Hydrogen Exchange Effect in the MgH2-LiBH4 System. J. Phys. Chem. C. 2010;114:13132–13135. DOI: 10.1021/Jp1042443

[28] Matsuo M, Nakamori Y, Orimo S, Maekawa H, Takamura H. Lithium superionic conduction in lithium borohydride accompanied by structural transition. Appl. Phys. Lett. 2007;91:224103. DOI: 10.1063/1.2817934

[29] Matsuo M, Orimo S. Lithium Fast-Ionic Conduction in Complex Hydrides: Review and Prospects. Adv. Energy Mater. 2011;1:161–172. DOI: 10.1002/aenm.201000012

[30] Unemoto A, Matsuo M, Orimo S. Complex Hydrides for Electrochemical Energy Storage. Adv. Funct. Mater. 2014;24:2267–2279. DOI: 10.1002/adfm.201303147

[31] Li HQ, Zhou HS. Enhancing the performances of Li-ion batteries by carbon-coating: present and future. Chem. Commun. 2012;48:1201–1217. DOI: 10.1039/C1cc14764a

[32] Maekawa H, Matsuo M, Takamura H, Ando M, Noda Y, Karahashi T, Orimo SI. Halide-Stabilized LiBH4, a Room-Temperature Lithium Fast-Ion Conductor. J. Am. Chem. Soc. 2009;131:894–895. DOI: 10.1021/Ja807392k

[33] Matsuo M, Remhof A, Martelli P, Caputo R, Ernst M, Miura Y, Sato T, Oguchi H, Maekawa H, Takamura H, Borgschulte A, Zuttel A, Orimo S. Complex Hydrides with (BH4)(-) and (NH2)(-) Anions as New Lithium Fast-Ion Conductors. J. Am. Chem. Soc. 2009;131:16389–16391. DOI: 10.1021/Ja907249p

[34] Oguchi H, Matsuo M, Sato T, Takamura H, Maekawa H, Kuwano H, Orimo S. Lithium–ion conduction in complex hydrides LiAlH4 and Li3AlH6. J. Appl. Phys. 2010;107:096104. DOI: 10.1063/1.3356981

Chapter 5

Intercalation of Poly[Oligo(Ethylene Glycol) Oxalate] into Vanadium Pentoxide Xerogel: Preparation, Characterization and Conductivity Properties

Evans A. Monyoncho, Rabin Bissessur,
Douglas C. Dahn and Victoria Trenton

Additional information is available at the end of the chapter

http://dx.doi.org/10.5772/62441

Abstract

We report, for the first time, the intercalation of poly[oligo(ethylene glycol) oxalate] (POEGO) and POEGO lithium salt ($LiCF_3SO_3$) complex (POEGO-$LiCF_3SO_3$) into vanadium pentoxide xerogel ($V_2O_5nH_2O$). The effect of changing the polymer concentration on the interlayer expansion of the layered host was studied, and the optimal intercalation ratio was determined to be 1:2. The intercalates were characterized by powder X-ray diffraction, thermogravimetric analysis, differential scanning calorimetry, Fourier transform infrared spectroscopy, and AC impedance spectroscopy.

Keywords: Layered structures, Nanocomposites, Electrical properties, Infrared (IR) spectroscopy, X-ray diffraction

1. Introduction

Intercalation chemistry is a versatile technique for developing new materials at moderate conditions. Unlike other techniques such as organic syntheses where rigorous conditions are used and purification of products is mandatory, intercalation chemistry provides an excellent route to combine the properties of two materials, which often does not require product purification. Therefore, intercalation chemistry is a "green shift" for developing new materials. Vanadium pentoxide xerogel ($V_2O_5nH_2O$) is a layered structure of interest to researchers in both academia and industry. This is because $V_2O_5nH_2O$ nanocomposites have potential applications such as in electrochemical energy storage devices [1, 2], biosensors [3], and electrochromic devices [4]. Most of the research on $V_2O_5nH_2O$ xerogel nanocomposites is focused in developing energy storage materials to address the current demands for renewable and portable electrical energy.

Intercalation of polymers into layered structures is a growing field of research with a wide range of potential applications [5]. For example, organic–inorganic nanocomposites offer promise for new engineering composites in the automotive, packaging, and aerospace industry because of their improved mechanical properties [6]. An organic–inorganic nanocomposite is a two-phase material in which the organic and inorganic phases are distributed in each other at the nanolevel. Therefore, with careful selection of the inorganic host and the guest polymer, it is possible to design materials that can be used in electrochemical energy storage devices such as in Li-ion batteries. Normally, the nanocomposite composition is controlled in order to increase its ionic conductivity at ambient temperatures so that it can be used as a solid electrolyte and/or as a cathode.

Vanadium pentoxide xerogel is a good host material for guest molecules and ions. The intercalation of guest species into vanadium pentoxide xerogel may occur via dipole–dipole interaction, ion exchange, acid–base, coordination, and redox reactions, enabling the system to accept both neutral and charged guest species [7, 8]. Vanadium pentoxide xerogel has also shown promising redox reactions that can be utilized in Li-ion batteries. For example, Passerini et al. demonstrated that $V_2O_5nH_2O$ xerogel could be used as a cathode material that reversibly intercalates more than three equivalents of lithium [9].

Current conventional Li-ion batteries utilize liquid organic electrolytes [10, 11] that come with several shortcomings, which limit their widespread usage in large load applications, such as electric vehicles and stationary power. Most of these liabilities are safety-related, which include electrolyte leakage, decomposition, flammability, and a propensity to develop catastrophic short circuits. To circumvent these problems, polymer electrolytes have been extensively studied as an alternative. The search for polymer electrolytes was initiated by the discovery of ionic conduction in complexes of poly(ethylene oxide) (PEO) containing alkali metal salts [12, 13] and the suggestion that such ionic conductors could be used as electrolytes in electrochemical devices [14]. Since then many PEO derivatives have been developed with efforts to improve ionic conductivity [15, 16]. There are two main strategies used for developing PEO derivatives with increased ionic conductivity: one approach focuses on increasing ionic mobility, that is, developing polymers that are flexible and amorphous with low glass transition temperatures. The second strategy focuses on increasing ionic dissociation by placing polar subunits such as acrylamide, acrylonitrile, maleic anhydride, and carbonate along the chains to increase the polymer's dielectric constant [17, 18]. Based on the second strategy, poly[oligo(ethylene glycol)-oxalate] (POEGO) was developed by Xu et al. [19]. They reported a maximum conductivity of 5.9×10^{-5} S cm^{-1} at 25°C with the complex of lithium bis(trifluoromethanesulfonyl)imide (LiTFSI), with etheric oxygen to Li-ion ratio of $[EO]/[Li^+] = 16$. They also reported that POEGO–LiTFSI complex showed good electrochemical stability up to 4.4 V versus Li^+/Li. These properties make POEGO a viable candidate for developing solid electrolytes for Li-ion batteries.

Several studies have been conducted where POEGO is intercalated into various layered structures. The intercalation is meant to improve the mechanical properties of POEGO in order to eliminate its tendency to flow under pressure, while retaining its ionic conductivity, and hence make it more suitable for use as a solid electrolyte in Li-ion batteries. For example, POEGO has been intercalated into hectorite [20], tin disulfide [21], graphite oxide [22], and molybdenum disulfide [23]. To our best knowledge, no work has been reported on the intercalation of POEGO into $V_2O_5nH_2O$ xerogel.

The focus of this chapter is to report on the intercalation of POEGO and $LiCF_3SO_3$–POEGO complex into vanadium pentoxide xerogel and the potentials of these nanocomposites in the Li-ion battery applications. In this chapter, the lithium salt ($LiCF_3SO_3$) and a complex of POEGO with $LiCF_3SO_3$ will be abbreviated as LiX ($X = CF_3SO_3$) and Li-POEGO, respectively. The nanocomposites will be abbreviated as V_2O_5POEGO and V_2O_5Li-POEGO followed by the mole ratio (e.g., V_2O_5POEGO 1:1).

2. Experimental

2.1. Materials

Sodium metavanadate was purchased from Fluka. Dowex 50W-X 8, 20–50 mesh resin, and oxalic acid dihydrate were purchased from Baker. HPLC-grade methanol was purchased from Caledon. Benzene 99.9% was purchased from Aldrich. All were used as received. Poly(ethylene glycol) (PEG 400), purchased from Aldrich was dried over 3 Å molecular sieves under nitrogen purge from a Schlenk line.

2.2. Vanadium pentoxide xerogel ($V_2O_5nH_2O$) synthesis

The synthesis of $V_2O_5nH_2O$ was adapted from the literature [24]. In this work, sodium metavanadate (4.0 g, 0.033 mol) was dissolved in deionized water (150 mL). The resulting colorless solution was then passed through an acidic ion exchange column (510 mEq capacity) made from Dowex 50W-X8, 20–50 mesh resin. The collected eluate (200 mL) was the yellow-brownish polyvanadic acid (HVO_3). The excess water was evaporated from the HVO_3 at room temperature for 2 weeks, and the solution was polymerized into a dark-red xerogel ($V_2O_5nH_2O$) solution. The $V_2O_5nH_2O$ solution was used without further dilution in the subsequent intercalation reactions.

2.3. Synthesis of poly[oligo(ethylene glycol) oxalate] (POEGO)

POEGO was synthesized as reported in the literature [19]. In a typical experiment, PEG 400 (15.0 g, 0.0375 mol) was added to oxalic acid dihydrate (5.00 g, 0.0397 mol) in benzene (100 mL), and the mixture was refluxed for 4 days in a 250 mL round-bottom flask while stirring. Benzene was then removed under reduced pressure. The reaction mixture was then heated in a vacuum oven at 120°C until a clear viscous product was obtained. The use of PEG with a molecular weight of 400 means that there are, on average, nine ethyleneoxy repeating units in each oligo(ethylene glycol) oxalate group in our POEGO [19].

A Li-POEGO complex was prepared using the optimum ionic conductivity ratio of $[EO]/Li^+$ = 16 as reported in the literature [19].

2.4. Synthesis of nanocomposites

The nanocomposites were prepared by adding POEGO dissolved in methanol to the aqueous solution of $V_2O_5nH_2O$, with stirring at room temperature in air for 30 min. In a typical experiment of mole ratio 1:1, 13.0 mL of the xerogel (0.160 g, 7.40×10^{-4} mol) was mixed with 1.0 mL of POEGO solution in methanol (1.25×10^{-4} mol). The following mole ratios were prepared: 1:0.5, 1:1, 1:2, 1:3, and 1:4. The mole ratios were calculated based on moles of $V_2O_51 \bullet 9H_2O$ and moles of POEGO's repeat unit. From here on, the nanocomposites will be referred to as V_2O_5POEGO followed by the mole ratio (e.g., V_2O_5POEGO 1:1 for the mole ratio of 1 to 1).

To confirm that the polymer chains were actually intercalated into $V_2O_5nH_2O$ and not just the solvent, a control sample was prepared and characterized. The control sample was prepared by adding 5.0 mL of methanol (the solvent used for dissolving POEGO) into 10.0 mL of an aqueous solution of $V_2O_5nH_2O$ gel, and stirring the mixture for 30 min. From here on, this control sample will be denoted as V_2O_5MeOH.

2.5. Materials characterization

Powder X-ray diffraction (XRD) data were collected on a Bruker AXS D8 Advance instrument equipped with a graphite monochromator, variable divergence slit, variable antiscatter slit, and a scintillation detector. Cu ($K\alpha$) radiation (λ = 1.5406 Å) was used. Samples were prepared as dry thin films on glass substrates by casting the homogeneous solutions at room temperature in air.

Thermal properties of the samples were investigated using TA instruments. Thermogravimetric analysis (TGA) data were collected on a Q500 in dry air or nitrogen purge using a heating rate of 10°C/min up to 680°C. Differential scanning calorimetry (DSC) was performed on a Q100 under dry nitrogen purge using heating and cooling rates of 10°C/min and 5°C/min, respectively. The TGA and DSC data were processed using the TA Universal Analysis 2000 software. The samples were freeze-dried using a Virtis Benchtop 3.3/Vacu-Freeze dryer and left in a desiccator overnight, before analysis.

Reflectance Fourier transform infrared (FTIR-ATR) spectra were obtained in the range 4000–400 cm^{-1} on a Bruker ALPHA FT-IR spectrometer equipped with attenuated total reflectance (ATR) sampling unit. The resolution of the instrument was 0.9 cm^{-1}, and 128 scans were used.

Conductivity data were collected by using AC impedance spectroscopy. A Solartron 1250 frequency response analyzer, along with a home-built current-preamplifier circuit was utilized. The amplitude of the sine wave perturbation was 50 mV, and a frequency range from 10 kHz to 0.01 Hz was used. The samples were run as cast films on rectangular glass substrates. Silver paste was placed on the two ends of the films as electrodes, so that current flow was along the film, parallel to the substrate. Prior to the conductivity measurements, samples were held in vacuum for at least 20 h at room temperature to remove any moisture and adsorbed volatile materials. During conductivity measurements, the samples remained in vacuum and were in thermal contact with an electrically heated copper sample holder cooled by a Cryodyne 350 refrigerator. A Lakeshore Cryotronics Model 321 temperature controller was used for temperature regulation. After the conductivity measurements, the glass substrates were cleaved so that the film thickness could be measured with an optical microscope.

3. Results and discussion

3.1. X-ray diffraction (XRD)

Powder XRD patterns were used to confirm the successful intercalation of POEGO and Li-POEGO into $V_2O_5nH_2O$ and the identity of the as-synthesized $V_2O_5nH_2O$. **Figure 1** shows the representative diffraction patterns for $V_2O_5nH_2O$, V_2O_5MeOH, and V_2O_5POEGO 1:1. The diffraction patterns of other nanocomposites showed similar features, and their interlayer expansions are summarized in **Table 1**. The interlayer spacing (*d*) for the synthesized $V_2O_5nH_2O$ was found to be 11.9 Å, and "*n*" was determined from TGA to be ≈1.9, which corroborates well with the literature values of $d = 11.5$ Å, $n = 1.8$ [25]. The *d*-spacing is dependent on the amount of intercalated water molecules. The diffraction pattern for V_2O_5POEGO 1:1 (**Figure 1**, curve c) showed a *d*-spacing of 20.4 Å, which corresponds to an interlayer expansion of 8.5 Å with respect to $V_2O_51.9H_2O$.

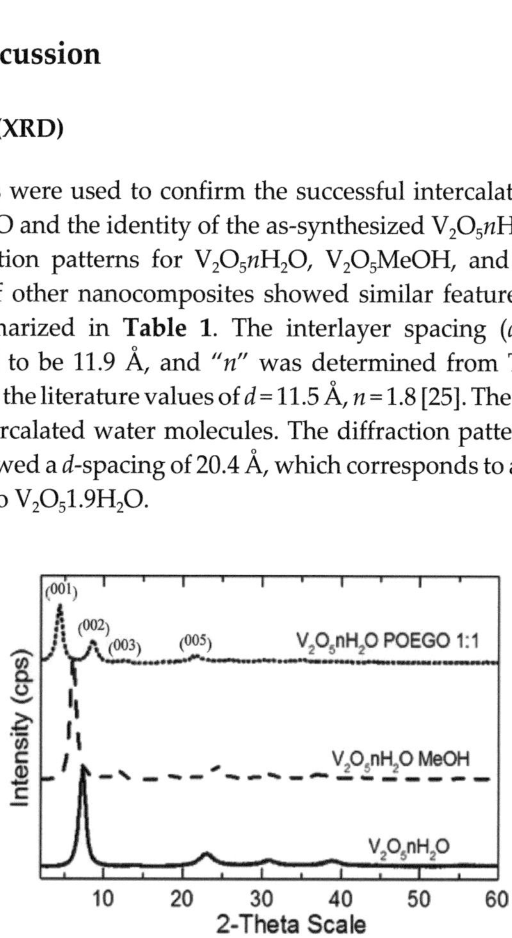

Figure 1. XRD of (a) $V_2O_5nH_2O$, (b) V_2O_5MeOH, and (c) V_2O_5POEGO 1:1.

Materials (ratios)	Observed *d*-spacing (Å)	Expansion Δ*d*-spacing (Å)	Average crystallite size (Å)
$V_2O_51.9H_2O$	11.9	–	73
V_2O_5MeOH	14.5	2.6	89

Materials (ratios)	Observed d-spacing (Å)	Expansion Δd-spacing (Å)	Average crystallite size (Å)
V_2O_5POEGO (1:0.5)	17.9	6.0	79
V_2O_5POEGO (1:1)	20.4	8.5	96
V_2O_5POEGO (1:2)	20.8	8.9	85
V_2O_5POEGO (1:3)	20.6	8.7	97
V_2O_5POEGO (1:4)	21.0	9.1	70

Table 1. List of interlayer expansions and average crystallite size of the nanocomposites.

The XRD patterns of the nanocomposites showed complete intercalation of POEGO in the gallery space of the host, based on the absence of the pristine $V_2O_5nH_2O$ phase. Further evidence of POEGO intercalation into $V_2O_5nH_2O$ is obtained through comparison of the XRD patterns to the control sample (V_2O_5MeOH), which showed a d-space increase of 2.6 Å compared to 8.5 Å for V_2O_5POEGO 1:1. These observations are strong evidence that POEGO was actually intercalated into $V_2O_5nH_2O$. However, it is difficult to say whether methanol is co-intercalated with POEGO from the XRD pattern alone. The XRD patterns of the nanocomposites show only (00l) reflections, which is indicative of highly oriented materials.

Table 1 shows that there is no significant increase in d-spacing as the amount of POEGO increases. This observation means that the average interlayer expansion of 8.7 Å corresponds to the optimum loading of POEGO chains into $V_2O_5nH_2O$. This finding corroborates with a previous work where POEGO has been intercalated into other layered structures. For example, intercalation of POEGO into tin disulfide [21] and molybdenum disulfide [23] were reported with interlayer expansions of 8.7 and 8.6 Å, respectively. From these literature reports, the interlayer expansion was ascribed to a bilayer arrangement of POEGO chains sandwiched between the layers of the host. Using similar arguments, the average interlayer expansion of 8.7 Å in this work corresponds to a bilayer conformation of POEGO chains between the V_2O_5 ribbons.

From the XRD patterns, the average crystallite size of the nanomaterials was calculated by using the Scherrer equation.[1] As shown in Table 1, there is no clear trend of the crystallite size with the amount of POEGO present in the nanocomposite. For example, the nanocomposite with mole ratio 1:3 shows the highest crystallite size, while that with mole ratio 1:4 shows the least value.

All the nanocomposite mole ratios prepared formed homogeneous solutions that were cast into thin films for XRD and conductivity studies. However, upon freeze-drying, the solutions from mole ratios 1:3 and 1:4 separated into two solid phases. One phase was a fine powder

[1] The Scherrer equation is $D_{hkl} = \frac{K\lambda 57.3}{\beta cos\theta}$, where D_{hkl} = average crystallite size (Å), λ is the wavelength of the Cu (K_α) radiation used (λ = 1.5406 Å), β = peak width at half height (2θ), and θ = position of the peak in degrees. The constant 57.3 is the conversion factor from radians to degrees. K is a constant that depends on the shape of the crystallites. The shape of our crystallites is not known. However, we use K = 0.9 (for spheres) for all samples, since we are primarily concerned with the trends, rather than the actual values.

and the other phase was a sticky mass. The two phases were characterized separately with the other techniques (TGA, DSC, and FTIR).

3.2. Thermogravimetric analysis (TGA)

The thermostability and stoichiometric composition of the nanocomposites were determined using TGA data. **Figure 2** shows the decomposition profile for POEGO, V_2O_5MeOH, V_2O_5POEGO 1:1, and V_2O_5POEGO 1:4 (the fine powder and the sticky mass). The thermograms were obtained in air using a heating rate of 10°C min^{-1}. POEGO decomposed completely in air, as shown in **Figure 2**, curve a. The control sample (V_2O_5MeOH) has the highest residue percentage of 89% as shown in **Figure 2**, curve e. Curves b and c of **Figure 2** show the decomposition profiles for V_2O_5POEGO 1:4 (sticky mass) and V_2O_5POEGO 1:4 (fine powder), respectively, while that of V_2O_5POEGO 1:1 is depicted by curve d. A complete analysis of the TGA data is summarized in **Table 2**. The nanocomposites have a weight residue percentage that corresponds to the amount of $V_2O_5nH_2O$ present.

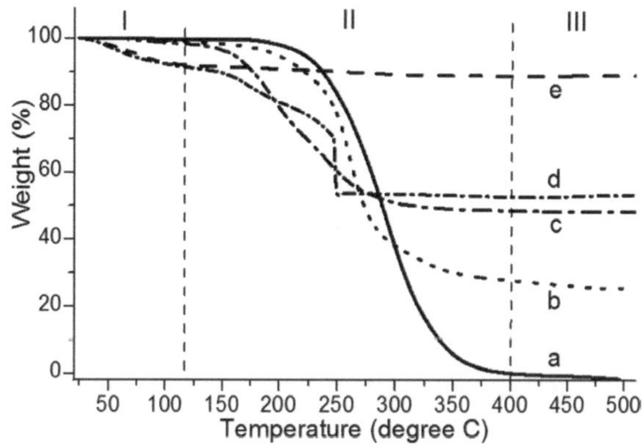

Figure 2. TGA of (a) POEGO, (b) V2O5POEGO 1:4 sticky mass (c) V2O5POEGO 1:4 powder, (d) V2O5POEGO 1:1, and (e) V2O5MeOH.

Material		Decomposition in air		Residue (wt.%)
Mole ratios used	Composition ratios	Weight (%)	Temperature (°C)[2]	
POEGO		100	291	0
$V_2O_5nH_2O$	$V_2O_5(H_2O)_{1.9}$	10.3, 2.5	49; (321, 364)	84.3
V_2O_5MeOH	$V_2O_5(MeOH)_{0.7}$	8.4	80	88.8
V_2O_5POEGO (1:0.5)	$V_2O_5(POEGO)_{0.1}(H_2O)_{0.8}$	22.3	285	72.2
V_2O_5POEGO (1:1)	$V_2O_5(POEGO)_{0.3}(H_2O)_{1.8}$	36.8	248	53.5
V_2O_5POEGO (1:2)	$V_2O_5(POEGO)_{0.6}(H_2O)_1$	55.7	204	40.2
V_2O_5POEGO (1:3) f	$V_2O_5(POEGO)_{0.6}(H_2O)_{0.7}$	59.0	(205, 254)	38.1

Material		Decomposition in air		Residue (wt.%)
Mole ratios used	Composition ratios	Weight (%)	Temperature (°C)[2]	
V_2O_5POEGO (1:3) s	$V_2O_5(POEGO)_{1.1}(H_2O)_{3.2}$	68.2	209	24.1
V_2O_5POEGO (1:4) f	$V_2O_5(POEGO)_{0.4}(H_2O)_{0.3}$	49.6	(195, 245)	48.6
V_2O_5POEGO (1:4) s	$V_2O_5(POEGO)_{1.2}(H_2O)_{0.6}$	72.9	260	25.5

Note. "f" denotes the fine powder phase and "s" denotes the sticky mass phase.

Table 2. TGA data for POEGO, $V_2O_5nH_2O$, V_2O_5MeOH, and the nanocomposites.

The thermogram of the control sample (**Figure 2**, curve e) shows a constant weight residue above 100°C, that is, after the evaporation of methanol. Therefore, the weight loss process above 100°C for the nanocomposites is associated with the polymer decomposition (see **Figure 2**). Increasing the amount of POEGO in the nanocomposites decreases the residue percentage, due to the decreased concentration of $V_2O_5nH_2O$, as shown in Table 2.

For the nanocomposites which separated into two phases, the fine powder phases (**Figure 2**, curve c) yielded higher residue percentages compared to the corresponding sticky mass phases (**Figure 2**, curve b). These data mean that the sticky phases have higher polymer component than the fine powder. Interestingly, the residue percentages of the sticky phases have similar weight percentages of 24.1% and 25.5% for V_2O_5POEGO 1:3 and V_2O_5POEGO 1:4, respectively. On the other hand, the corresponding residue percentages of the fine powders are significantly different, having a difference of more than 10%. This observation means that during the phase separation, the composition of the sticky phase is independent of the amount of $V_2O_5nH_2O$ or POEGO used.

The decomposition profiles can be divided into three stages as shown in **Figure 2**. In stage I (<120°C), the weight loss corresponds to the evaporation of water/solvent. The weight loss in stage I varies randomly with no correlation to the amount of polymer used. In stage II (120–400°C), the weight loss corresponds to the decomposition of the polymer. At this stage, the percentage weight loss is directly proportional to the amount of the polymer present in the nanocomposite. Finally, stage III (>400°C) corresponds to the residue which has a constant weight percentage. The residue, yellow in color, was identified with XRD to be orthorhombic V_2O_5 crystals.

The stoichiometric compositions were calculated based on the mass loss at each of the first two stages, the mass of the residue, and the corresponding molecular weight of the compound.[3] A complete list of calculated compositions for all nanocomposites is provided in Table 2. There is no clear correlation (from Table 2) between the experimental mole ratios and the calculated

[2] The temperature values were taken from the peak maximum of the derivative plot. Where the derivative peaks were not well resolved (due to overlapping weight loss steps), the temperature values of the peaks are enclosed in brackets.

[3] The mass loss at stage I, stage II, and the mass of the residue (stage III) were assumed to correspond to water, POEGO, and V_2O_5, respectively. The moles of water, POEGO, and V_2O_5 were calculated for each sample. The mole ratios were expressed with respect to the moles of V_2O_5 (e.g., the composition for V_2O_5POEGO 1:1 was determined to be $V_2O_5(POEGO)_{0.3}(H_2O)_{1.8}$).

http://dx.doi.org/10.5772/62441

composition ratios from TGA. A possible explanation for this observation is due to the solid-phase separations, which was not controllable experimentally, thus affecting the homogeneity of the samples.

3.3. Differential scanning calorimetry (DSC)

DSC provided important information on *thermal transitions* occurring in the nanocomposites. **Figure 3** illustrates the results found for POEGO, V_2O_5POEGO 1:1, V_2O_5POEGO 1:4 fine powder phase, and V_2O_5POEGO 1:4 sticky phase, respectively. The significant transitions observed for these nanocomposites were the glass transition temperature (T_g) and the blending temperature (T_e). Other nanocomposites show similar thermal transitions. A complete list of glass transitions and blending temperatures are summarized in Table 3. The glass transition temperature for POEGO is close to the value of −55.0°C previously reported for POEGO of similar composition (see [19], Table 3, $n = 9$).

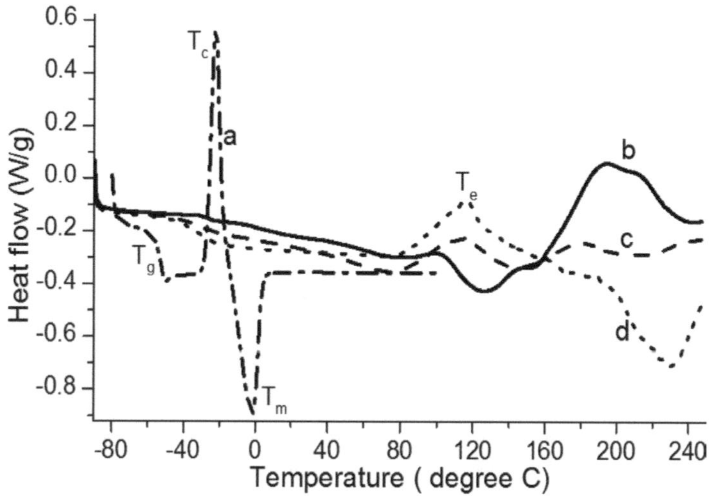

Figure 3. DSC of (a) POEGO, (b) V2O5POEGO 1:1, (c) V2O5POEGO 1:4 fine powder, and (d) V2O5POEGO 1:4 sticky mass at a scan rate of 10°C min^{-1} (T_c is the recrystallization temperature, and T_m is the melting temperature).

Materials (mole ratio)	Phase transitions		ΔT (T_e–T_g)
	T_g (°C)	T_e (°C)	
POEGO	−52.2	–	–
V_2O_5POEGO (1:1)	−25.9	99	124.9
V_2O_5POEGO (1:2)	−25.2	108	133.2
V_2O_5POEGO (1:3) f	−30.4	115	145.4
V_2O_5POEGO (1:3) s	−35.3	113	148.3

Materials (mole ratio)	Phase transitions		ΔT (T_e–T_g)
	T_g (°C)	T_e (°C)	
V_2O_5POEGO (1:4) f	–30.0	115	145.0
V_2O_5POEGO (1:4) s	–31.6	116	147.6

Note. "f" denotes the fine powder phase and "s" denotes the sticky mass phase.

Table 3. DSC thermal transitions for POEGO and nanocomposites.

All samples show negative sloping baselines after the glass transition temperature, which indicates the slow heat flow to the sample, which may correspond to the energy used to vaporize volatiles from the samples during heating. The horizontal baseline displayed by the sticky phase (**Figure 3**, curve d) means that the sample and the reference cell are in thermodynamic equilibrium. This observation is characteristic for amorphous and flexible materials [26].

The blending temperature being below the decomposition temperature (<200°C) means that thin films can be obtained by hot pressing the fine powder or the sticky mass up to 100°C, without any decomposition or damage.

The DSC comparison between POEGO (**Figure 3**, curve a) and V_2O_5POEGO 1:4 sticky mass phase (**Figure 3**, curve d) showed that the sticky mass phase did not have any crystallization or melting *thermal transitions* between the glass transition and the blending temperatures. This observation means that the sticky mass phases are amorphous and flexible between the glass transition and the blending temperatures. This is a significant finding because amorphous polymers are known to be better ionic conductors than their crystalline counterparts. In addition, the stickiness improves the binding ability of the materials, making the nanocomposites good candidates for solid electrolyte and/or cathode materials for lithium/Li-ion batteries.

The temperature gap (ΔT) between T_g and T_e was calculated for each sample and is shown in Table 3. The glass transition temperatures of the nanocomposites are higher than that of pure POEGO (Table 3), because the nanocomposites are more rigid than POEGO, and are hence easily turned into the glassy state.

3.4. Fourier transform infrared (FTIR)

FTIR spectroscopy was used to investigate the type of chemical bonds present in the nanocomposites in comparison to POEGO and pristine $V_2O_5nH_2O$. **Figure 4** shows the IR spectra for POEGO, $V_2O_5nH_2O$, V_2O_5POEGO 1:1, and V_2O_5POEGO 1:4 sticky mass, respectively. IR spectra for all nanocomposites exhibit similar features. The IR band assignments were as follows: POEGO ν_{max}/cm^{-1} 2872w, 952 and 861 (CH), 1770s and 1745s

(COCO), and 1115s (C–O–C); $V_2O_5nH_2O$ ν_{max}/cm^{-1} 1616 (HOH), 999 (V=O), and 707 (V–O–V); and a typical nanocomposite V_2O_5POEGO 1:1 ν_{max}/cm^{-1} 2867 and 941 (CH), 1623 (COCO), 1082, 1021, and 986 (C–O–C and V=O).

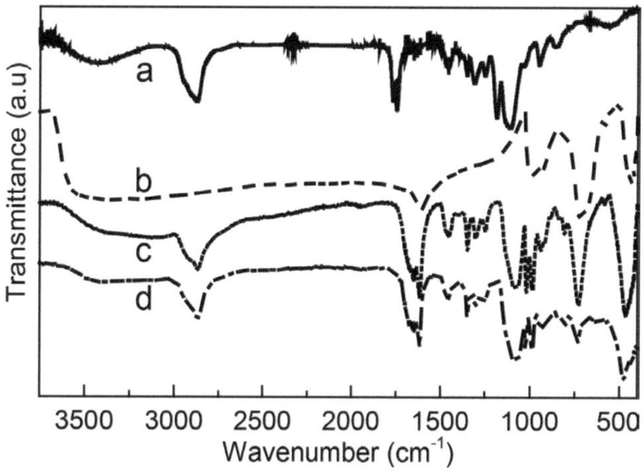

Figure 4 FTIR spectrum for (a) POEGO, (b) $V_2O_5nH_2O$, (c) V_2O_5POEGO 1:1, and (d) V_2O_5POEGO 1:4 sticky mass.

It is important to note that comparison of the IR spectra for $V_2O_5nH_2O$ among various literature reports is complicated due to the distinct structural differences associated with the methods of $V_2O_5nH_2O$ preparation and the aging period of the precursor solutions. In general [27], the characteristic V=O stretching absorption bands are found within the 950–1020 cm^{-1} range, while the bridging V–O–V stretching absorption bands usually occur between 700 and 900 cm^{-1}. The IR bands below 600 cm^{-1} correspond to either edge-sharing V–O stretching or bridging V–O–V deformation vibrations.

The absorption bands for C=O and V=O were significantly perturbed in the nanocomposites, an indication for a chemical reaction between POEGO and $V_2O_5nH_2O$ during intercalation. However, the presence of bands at 2867, 1082, 986, and 1021 cm^{-1} confirms that the structure of the parent polymer and that of the layered structure are retained to some degree in the nanocomposites. It is not conclusive as to what kind of new chemical bonds are formed. The C–H stretching vibrations for POEGO shifted from 2872 cm^{-1} to between 2867 and 2845 cm^{-1} in the nanocomposites. This band shift provides evidence of an interaction between the polymer and the layered structure.

3.5. AC impedance spectroscopy

The impedance experiment involved applying an AC voltage to the sample and measuring the real and imaginary parts of the resulting current. Curve (a) in **Figure 5** is a complex plane plot of the impedance of a Li-POEGO cast-film sample at a temperature of 310 K. High-frequency data is near the origin. A semicircle like this, with a low-frequency diagonal spur on the right, is characteristic of an ionic conductor between blocking electrodes [28]. Also shown is a fit to the equivalent-circuit model shown in the inset in **Figure 5**. In this circuit, *R* is the ionic resistance of the sample, *C* includes the sample capacitance but is dominated by cable capacitance in parallel with it, and the constant phase element CPE models electrode effects. From the fit, we get $R = (4.43 \pm .04) \times 10^7\ \Omega$. The sample had a length between electrodes of 9.2 mm, a width of 8.6 mm, and a thickness of (50 ± 10) μm. The ionic conductivity calculated from *R* and the sample dimensions is $(5 \pm 1) \times 10^{-6}$ S/cm. This is within the range previously reported for Li-POEGO [19], but is less than the earlier data for Li-POEGO with nine repeat units in the oligomers (see, e.g., [19], **Figure 4**, $n = 9$). The difference may be related to different sample preparation techniques; the earlier measurements were made using dip-type cells, while our samples are cast films.

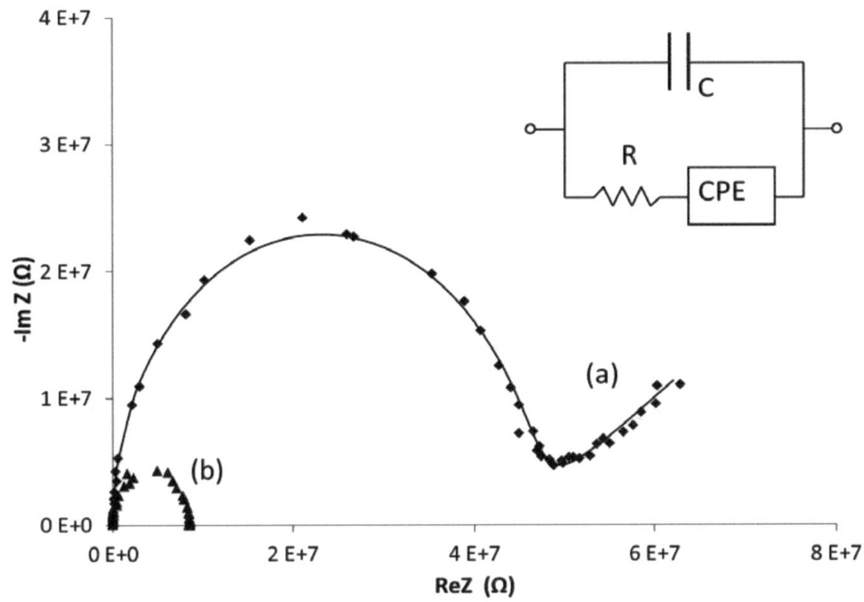

Figure 5 (a) Complex plane plot of the impedance of Li-POEGO at 310 K. The line is a fit to the equivalent circuit shown in the inset. (b) Impedance data for V2O5POEGO 1:4 at 300 K.

Impedance data for a typical nanocomposite, V_2O_5POEGO 1:4 (batch EM1-77), at 300 K, is also shown in **Figure 5** (curve b). The plots for all our nanocomposites are semicircles, similar to the plots for a simple parallel RC circuit, indicating that the nanocomposites are electronic conductors. Here, *R* is the resistance of the sample and *C* is the total capacitance in parallel with it. *C* includes the cable capacitance and the capacitance due to the interfaces between the silver paste electrodes and the V_2O_5POEGO sample. We find experimentally that when *C* is determined by fitting impedance data on these samples, the value of *C* is equal, within experimental error, to the cable capacitance alone (about 310 pF in our system). This indicates that other contributions to *C* are very small in comparison. The value of *R* can be determined

http://dx.doi.org/10.5772/62441

directly from the complex-plane plot, since the low-frequency end of the semicircle terminates on the real axis at R (frequency decreases from left to right). Conductivity (σ) is then calculated from R and the dimensions of the sample. For example, the sample shown in **Figure 5(b)** had $R = (8.52 \pm 0.05) \times 10^6\ \Omega$, length between electrodes of 6.9 mm, a width of 9.5 mm, and a thickness of (18 ± 6) μm. The conductivity is therefore $(4.7 \pm 1.6) \times 10^{-5}$ S/cm. The large uncertainty in the thickness reflects the nonuniform thickness of the cast films. This is the main source of uncertainty in all of our conductivity values.

Electronic conductivity data for $V_2O_5nH_2O$ xerogel, and for a number of nanocomposite samples, are shown as a function of inverse temperature in **Figure 6**. The conductivity of $V_2O_5nH_2O$ (3.0×10^{-3} S/cm at 300 K) is comparable to the values reported in the literature. Reported conductivity values for $V_2O_5nH_2O$ vary widely, however, due to many variable experimental parameters, such as the state of vanadium reduction, the relative humidity of the atmosphere, the age of the gel, the sample film dimensions, and the direction of conductivity measurements. All these parameters affect the conductivity of $V_2O_5nH_2O$. For example, due to the layered structure of $V_2O_5nH_2O$, its conductivity can be four orders of magnitude larger when measurements are performed in a direction parallel to the ribbons rather than perpendicular to the ribbons [29]. The conductivities reported in this work were measured approximately parallel to the ribbons, that is, the current flow was along the film, parallel to the substrate.

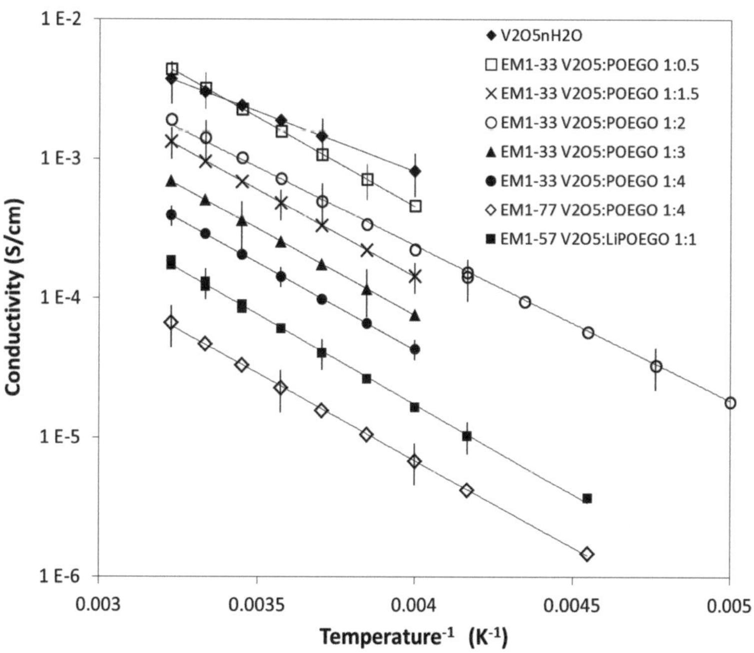

Figure 6 Temperature-dependent conductivities of $V_2O_5nH_2O$, and V_2O_5POEGO nanocomposites. To avoid clutter, error bars are shown only for a few representative points. Uncertainties for other points are similar.

The results for the nanocomposite samples from sample group EM1-33, which were all made at the same time, using the same $V_2O_5nH_2O$ and POEGO starting materials, clearly show the general trend that conductivity decreases as the amount of intercalated POEGO increases. This is consistent with the fact that POEGO is an electrical insulator. Samples made from different batches of starting materials showed a similar trend, but the conductivity values were not consistent from one batch to the next. One example is shown in the **figure 6**: the V_2O_5POEGO 1:4 nanocomposite from batch EM1-77 had a conductivity more than six times lower than that of the V_2O_5POEGO 1:4 nanocomposite from batch EM1-33. This variability between batches is not surprising, given the variability in the conductivity of $V_2O_5nH_2O$ mentioned earlier.

Conductivities of a few V_2O_5LiPOEGO nanocomposites were also measured, and one typical sample is included in **Figure 6**. In general, the electrical conductivity of V_2O_5LiPOEGO was less than that of V_2O_5POEGO. We were not able to detect any signs of ionic conductivity in the V_2O_5LiPOEGO nanocomposites. However, with the impedance technique used in this work, if the ionic conductivity is much smaller than the electronic conductivity, it will not be detected. A material like V_2O_5LiPOEGO 1:4 is expected to conduct lithium ions, and these experiments do not rule that out, since as long as the ionic conductivity is below about 10^{-5} S/cm at 300 K, we would not be able to observe it.

When conductivity data plot as straight lines in an Arrhenius-type plot like **Figure 6**, this indicates a thermally activated conduction process described by an equation of the form

$$\sigma(T) = \sigma_0 e^{-E/kT}$$

where σ_0 is a constant and E is the activation energy (or semiconductor band gap). Fits to this equation are shown as solid lines in **Figure 6**, and the activation energies were found from the slopes of these lines. For $V_2O_5nH_2O$, we get $E = 0.17$ eV, close to the values reported in the literature [29, 30]. All the nanocomposites have activation energies in the range 0.23–0.26 eV, very close to each other and significantly higher than that for $V_2O_5nH_2O$. This can be seen directly in the graph where the nanocomposite lines all have roughly the same slope, and all have a steeper slope than $V_2O_5nH_2O$. We conclude that insertion of PEOGO into vanadium pentoxide increases the band gap. A similar effect was seen when poly(oxymethylene-oxyethylene) (POMOE) was inserted into $V_2O_5nH_2O$ [31].

Note that the band gap does not depend on the V_2O_5:POEGO ratio. Our XRD data shows that interlayer expansion is also almost independent of V_2O_5:POEGO ratio, indicating that the interlayer spaces are fully occupied with a bilayer arrangement of POEGO chains even before the ratio reaches 1:1. Since increasing the amount of POEGO further does not result in more POEGO between the V_2O_5 layers, some of the POEGO must be outside the nanocomposite crystallites. This interpretation of the XRD data explains two features of the conductivity. First, assuming the increase in activation energy is due to separation of the V_2O_5 layers and/or interaction between the V_2O_5 layers and the inserted polymers, there should be no further change in activation energy once the interlayer spaces are full. Second, as the amount of POEGO is increased beyond what is needed to fill the interlayer spaces, regions of electrically insulating polymer will form outside the nanocomposite crystallites, blocking some charge-transport channels and reducing the conductivity through the sample.

Since the conductivity measurements were made with cast films in which phase separation does not occur, we have not yet acquired the conductivity data on the separate fine powder and sticky mass phases obtained from freeze-dried V_2O_5POEGO 1:3 and V_2O_5POEGO 1:4. Characterization of charge transport in these phases would be an interesting topic for future work. Because of the high polymer content in the sticky mass phase, complexing it with a lithium salt may yield a material with higher ionic conductivity.

4. Conclusions

In summary, successful intercalation of POEGO and Li-POEGO into $V_2O_5nH_2O$ at room temperature is demonstrated. The sticky mass phases of the nanocomposites obtained upon freeze-drying the mole ratios above 1:3 have appealing properties, as they are thermally stable in air up to around 200°C and amorphous in the temperature range (−30°C to 100°C). The cast film nanocomposites we studied using AC impedance spectroscopy are not suitable for solid electrolyte applications because of their electronic conductivity. V_2O_5LiPOEGO nanocomposites prepared using cast film methods are believed to be conductors of both electrons and lithium ions, and therefore may have applications as electrode materials in Li-ion batteries. The next challenge is to further characterize the phase-separated materials produced by freeze-drying.

Acknowledgements

The authors are grateful for the financial support from the Natural Sciences and Engineering Research Council (NSERC) of Canada, Canada Foundation for Innovation (CFI), Atlantic Innovation Fund of Canada (AIF), Innovation Prince Edward Island (PEI), and University of Prince Edward Island (UPEI).

Author details

Evans A. Monyoncho[1], Rabin Bissessur[1*], Douglas C. Dahn[2] and Victoria Trenton[2]

*Address all correspondence to: rabissessur@upei.ca

1 Chemistry Department, University of Prince Edward Island, Charlottetown, PE, Canada

2 Physics Department, University of Prince Edward Island, Charlottetown, PE, Canada

References

[1] Yang Z, Zhang J, Kintner-Meyer MCW, Lu X, Choi D, Lemmon JP, Liu J. Electrochemical energy storage for green grid. *Chem Rev* 2011;111:3577–3613.

[2] Baddour R, Pereira-Ramos JP, Messina R, Perichon J. Vanadium pentoxide xerogel as rechargeable cathodic material for lithium batteries. *J Electroanal Chem* 1990;277:359–366.

[3] Tsiafoulis CG, Florou AB, Trikalitis PN, Bakas T, Prodromidis MI. Electrochemical study of ferrocene intercalated vanadium pentoxide xerogel/polyvinyl alcohol composite films: application in the development of amperometric biosensors. *Electrochem Commun* 2005;7:781–88.

[4] Toma HE. Vanadium(V) oxide—metal organic nanocomposites as electrochemical sensing materials. *Mater Sci Forum* 2010;636:729.

[5] Müller-Warmuth W, Schöllhorn R. *Progress in Intercalation Research.* In *Physics and Chemistry of Materials with Low-dimension Structures.* Kluwer Academic Publishers, USA, 1994, Vol. 17, pp. 536.

[6] Pomogailo AD. Synthesis and intercalation chemistry of hybrid organic and inorganic nanocomposites. *Polym Sci* 2006;48:85–111.

[7] Zakharova GS, Volkov VL. Intercalation compounds based on vanadium(V) oxide xerogel. *Russ Chem Rev* 2003;72:311–325.

[8] Wu CG, Kanatzidis MG. Layered vanadium pentoxide xerogels: host-guest chemistry and conductive-polymers. *Mater Res Soc Symp Proc* 1991;210:429–442.

[9] Passerini S, Chang D, Chu X, Ba Le D, Smyrl W. Spin-coated V_2O_5 xerogel thin films. 1. Microstructure and morphology. *Chem Mater* 1995;7:780–85.

[10] Ramzy A, Thangadurai V. Tailor-made development of fast Li Ion conducting garnet-like solid electrolytes. *ACS Appl Mater Interfaces* 2010;2:385–390.

[11] Dietrich Berndt. *Batteries,* 3. Secondary Batteries. In *Ullmann's Encyclopedia of Industrial Chemistry.* Wiley-VCH Verlag GmbH & Co. KGaA, 2014, 1–68.

[12] Fenton DE, Parker JM, Wright PV. Complexes of alkali metal ions with poly(ethylene oxide). *Polymer* 1973;14:589.

[13] Wright PV. Electrical conductivity in ionic complexes of poly(ethylene oxide). *Br Polym J* 1975;7:319–327.

[14] Armand MB, Chabagno JM, Duclot M. In *Poly-ethers as solid electrolytes.* Vashitshta P, Mundy JN, Shenoy GK, editors, Second International Meeting on Solid Electrolytes, North Holland Publishers, 1978. Amsterdam, Netherlands.

[15] Quartarone E, Mustarelli P, Magistris A. PEO-based composite polymer electrolytes. *Solid State Ionics* 1998;110:1–14.

[16] MacCallum JR. In *Polymer Electrolyte Reviews*, MacCallum JR and Vincent CA, editors, Elsevier Applied Science, 1987. London and New York.

[17] Forsyth M, Tipton AL, Shriver DF, Ratner MA, Mac Farlane DR. Ionic conductivity in poly(diethylene glycol-carbonate)/sodium triflate complexes. *Solid State Ionics* 1997;99:257–261.

[18] Xu K, Zhou T, Deng ZH, WanGX. *Chin J Polym Sci* 1992;10:223.

[19] Xu W, Belieres J, Angell A. Ionic conductivity and electrochemical stability of poly[oligo(ethylene glycol)oxalate]-lithium salt complexes. *Chem Mater* 2001;13:575–380.

[20] Scully SF, Bissessur R. Encapsulation of polymer electrolytes into hectorite. *Appl Clay Sci* 2010; 47:444–447.

[21] Bissessur R, Schipper D. Exfoliation and reconstruction of Sn S_2 layers: a synthetic route for the preparation of polymer-Sn S_2 nanomaterials. *Mater Lett* 2008;62:1638–1641.

[22] Bissessur R, ScullySF. Intercalation of solid polymer electrolytes into graphite oxide. *Solid State Ionics* 2007;178:877–882.

[23] Bissessur R, Gallant D, Brüning R. Novel Intercalation compound of poly[oligo(ethylene glycol)-oxalate in molybdenum disulfide. *J Mater Sci Lett* 2003;22:429–431.

[24] Guerra EM, Ciuffi KJ, Oliveira HP. V_2O_5 xerogel–poly(ethylene oxide) hybrid material: synthesis, characterization, and electrochemical properties. *J Solid State Chem* 2006;179:3814–3823.

[25] Livage J. Synthesis of polyoxovanadates via "chimie douce". *Coord Chem Rev* 1998:178/180:999–1018.

[26] Wunderlich B. *Thermal Analysis of Polymeric Materials*. Springer-Verlag, Berlin; Heidelberg, Berlin, Heidelberg, 2005.

[27] Anaissi FJ, Demets GJ, Alvarez EB, Politi MJ, Toma H.E. Long-term aging of vanadium(V) oxide xerogel precursor solutions: structural and electrochemical implications. *Electrochim Acta* 2001;47:441–450.

[28] Barsoukov E. In Macdonald JR, editors. *Impedance Spectroscopy: Theory, Experiment, and Applications*. John Wiley & Sons, Hoboken, New Jersey, USA, 2005.

[29] Livage J. Vanadium pentoxide gels. *Chem Mater* 1991;3:578–593.

[30] De S, Dey A, De SK. Characterization and transport properties of intercalated polypyrrole- vanadium pentoxide xerogel nanocomposite. *Solid State Commun* 2006;137:662–667.

[31] Monyoncho E, Bissessur R, Trenton V, Dahn DC. A bilayer insertion of poly(oxymethylene-oxyethylene) into vanadium oxide xerogel: preparation, characterization and insertion mechanism. *Solid State Ionics* 2012;227:1–9.

Chapter 6

Highly Functionalized Lithium-Ion Battery

Hiroki Nagai and Mitsunobu Sato

Additional information is available at the end of the chapter

http://dx.doi.org/ 10.5772/63491

Abstract

Future energy demand is an important issue that requires consideration. Lithium-ion batteries (LIB) are one of the most popular types of rechargeable battery for portable electronic devices, such as mobile phones, cameras, and laptop computers, and have led to other applications being commercialized. Distributed power generation using renewable energy sources, such as solar photovoltaic (PV), can efficiently supply electricity according to on-site demand. If the electrodes of a LIB could function as a solar cell, the storage device could provide electricity without an electric power supply.

We have reported the fabrication of a thin-film LIB using the molecular precursor method. The thin-film LIB was fabricated with $Li_4Ti_5O_{12}$ for the anode and $Li_3Fe_2(PO_4)_3$ for the cathode on a fluorine-doped tin oxide pre-coated glass substrate and an electrolyte of dissolved $LiPF_6$. Moreover, we fabricated a novel LIB that was charged by light. This novel, translucent, thin-film LIB that can be charged by solar light irradiation was fabricated using active materials—titania for the anode and $LiCoO_2$ for the cathode—on a conductive glass substrate by a wet process.

This chapter focuses on the fabrication and properties of these multi-functional thin-film LIBs using a chemical method, viz., the molecular precursor method. The translucent device can simultaneously generate and store electricity from solar light and may be applied in smart windows, facilitating the wider use of renewable energy.

Keywords: lithium-ion battery, thin film, translucent, molecular precursor method, light irradiation

1. Introduction

Resource scarcity, energy, and global warming are all problems faced by modern society. In recent years, access to energy has become a particularly acute problem. Energy prices have soared worldwide, and demand is increasing. One possible solution for these problems is the development of energy devices such as rechargeable batteries, solar cells, and fuel cells.

Lithium batteries using the Li metal or Li compounds as the anode have been studied over the past two decades; however, it has been the safety problems [1]. To overcome the problem, a safer approach has been to replace Li metal with a Li intercalation compound, usually a carbon compound, leading to the so-called "lithium ion" batteries (LIBs) [2–4]. The energy density of a LIB with high discharge voltage (3.6 V) is nearly twice that of Ni-Cd batteries, and excellent cycle life and a higher level of intrinsic safety have been demonstrated. Currently, the rechargeable LIB is one of the most popular types of rechargeable battery for portable electronic devices such as mobile phones, cameras, and laptop computers. Many other LIB applications have also been commercialized. In particular, micro-sized LIBs have attracted much interest for use as power sources for smart cards, implantable medical devices, and micro-sensors.

We recently fabricated a transparent thin-film LIB using the molecular precursor method (MPM) [5]. The device was fabricated with $Li_4Ti_5O_{12}$ (LTO) as the anode and $Li_3Fe_2(PO_4)_3$ (LFP) as the cathode on a fluorine-doped tin oxide (FTO) pre-coated glass substrate, with an electrolytic solution of $LiPF_6$. We discovered that this novel LIB is also electrochromic, that is, it colors and discolors according to charge and discharge operations, respectively. This type of electrochromic device can be used in various monitoring systems, electrically controllable tint glass, and so on. If the electrodes of a LIB could also function as a solar cell, the storage device could then provide electricity without an electric power supply.

2. Principle of the molecular precursor method

The MPM is one of the wet processes used for thin film fabrication of various metal oxides, including metal oxides or calcium phosphate compounds [6]. This method is based on the design of metal complexes in coating solutions with excellent stability, homogeneity, miscibility, and coatability, and so on. These highly stable metal-complex anions are highly soluble in water and can thus be dissolved in volatile solvents by combining them with appropriate alkylamines. Furthermore, the resultant solutions can generally form excellent precursor films using various coating procedures such as spin coating. The precursor films produced from metal complexes should be amorphous, similar to films produced by metal/organic polymers in sol-gel processes; otherwise, it would not be possible to fabricate homogeneous metal-oxide thin films on substrates using heat treatment methods.

A stable metal complex is dissolved at a molecular level in the precursor solution. The metal complex salt in the precursor film must be amorphous before heat treatment to fabricate thin films without cracks and pinholes. The alkylamines play an important role in obtaining an amorphous salt in the precursor film. The plausible packing of the metal complex in the precursor film formed on the substrate can be theoretically explained using molecular dynamics. The shrinkage rate of the film in the vertical direction can be estimated from the model structure before heat-treatment based on the crystal structure of the metal complex salt, which can be obtained as a single crystal when the alkyl groups in the amines are short enough.

The shrinkage rate in the sol-gel method is usually considered to be around 10 times. However, it is roughly estimated to be 10–15 times in the case of MPM, on the basis of the crystal structures (**Figure 1**). Thus, the densification degrees of the precursor films during heat treatment in the process of MPM are similar to those from sol-gel procedures, even though the precursor films involve alkylamines and ligands.

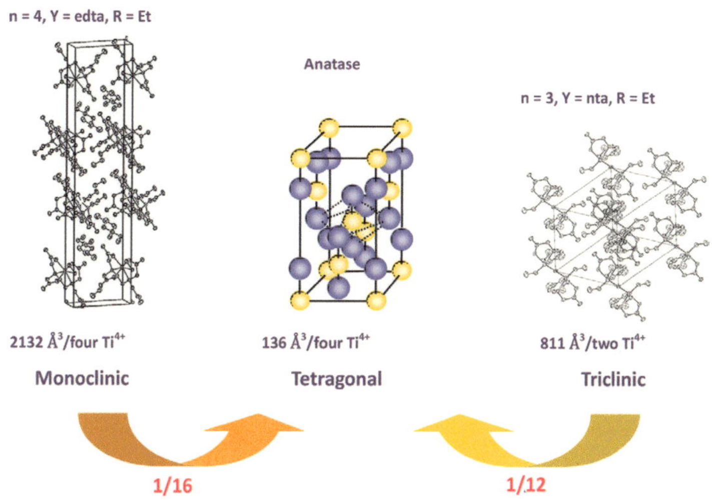

Figure 1. Schematic for shrinkage models of metal complex film (left side: Ti-edta complex; right side: Ti-nta complex).

3. Electrochromism of the transparent LIB

Recently, 3-D structures formed from compounds such as PO_4 tetrahedra and FeO_6 octahedra have created interest for their potential use as cheap positive electrodes for lithium-ion batteries. Efforts toward this relatively novel class of intercalation hosts for lithium have focused on systems such as the olivine $Li_{1-x}FePO_4$ and NASICON compositions $Li_xFe_2(SO_4)_3$ or $Li_3Fe_2(PO_4)_3$ (LFP) into which reduction/oxidation of Fe^{3+}/Fe^{2+} occurs at potentials close to 3.43, 3.55, and 2.8 V vs. Li/Li^+, respectively [7–10].

Spinel $Li_4Ti_5O_{12}$ (LTO) is a promising anode material for LIBs. This material exhibits very flat discharge and charge curves and only a small structural change during charge-discharge processes. The theoretical specific capacity of this material is 175 mA h g^{-1} and exhibits a practical specific capacity as high as 150–160 mA h g^{-1} after 100 discharge cycles. During this process of Li^+ intercalation and de-intercalation, the cubic symmetry of the parent spinel is Unaffected by lithiation; the lattice parameter (8.36 Å) and unit-cell volume are virtually unaltered by the phase transition. The lack of significant changes to the crystallographic parameters is remarkable and provides a structure that is extremely tolerant to electrochemical cycling. At the same time, the improved safety and reliability of the spinel compared with that of carbon electrodes make the LIBs using LTO as anode material suitable for electric vehicle (EV) and power storage batteries [11–15].

We attempted to fabricate a transparent thin-film LIB using the MPM. A transparent LFP thin-film cathode of 80 nm thickness was fabricated on a conductive FTO pre-coated glass substrate by heat treating a precursor ethanolic solution containing a Li(I) complex of nitrilotriacetic acid (NTA), an Fe(III) complex of ethylenediaminetetraacetic acid (EDTA), and $(dibutylammonium)_2H_2P_2O_7\cdot0.5H_2O$ at 550 °C for 10 min in air. A transparent LTO thin-film anode of 90 nm thickness was also fabricated on the substrate by heat treating a precursor ethanolic solution containing a Li(I) complex of NTA, a Ti(IV) complex of NTA, and hydrogen peroxide, at 550 °C for 30 min in air. The rechargeability of the assembled sandwich-type battery using an electrolytic solution dissolving $LiPF_6$ was measured by a repeated charge and discharge test.

The weight loss of 17% accompanied by an endothermic peak in the temperature range from 100 to 200°C appeared in the TG-DTA curves of the LTO precursor gel (**Figure 2**). The TG curve shows that the compound decomposed by 600°C with a final residual weight of 13%. The $Li_4Ti_5O_{12}$ powders are usually synthesized by a solid-state reaction of lithium and titanium salts [16, 17]. However, there are some disadvantages to solid-state reactions, such as inhomogeneous distribution, lack of stoichiometric control, and so on [18]. Hao and co-workers fabricated a spinel-type $Li_4Ti_5O_{12}$ by a modified sol-gel method using oxalic acid as a chelating agent. According to the TG-DTA results of the report, the final weight loss from 700 to 800°C is mainly attributed to the thermal decomposition of residual carbonate phases and the completion of the crystallization to $Li_4Ti_5O_{12}$. These results correspond to the following reaction:

$$2Li_2CO_3 + 5TiO_2 \rightarrow Li_4Ti_5O_{12} + 2CO_2$$

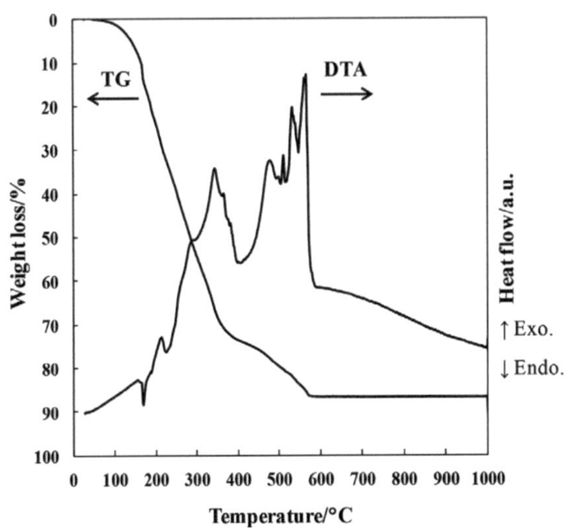

Figure 2. The TG-DTA curves of the LTO precursor gel obtained by evaporating the precursor solution. The measurement temperature was increased from 25 to 1000°C at a rate of 10°C min^{-1} using an air flow rate of 0.1 L min^{-1}.

However, in the case of MPM, no reaction occurs between 700 and 800°C from the TG-DTA results. Thus, it is significantly suggested that the chelating ligand in the molecular precursors successfully prevents any unexpected segregation of the metal compounds during oxide formation because the discrete molecular precursor complexes provide homogeneous and ideal mixtures of the different metal species at the molecular level in the precursor films.

The repeated charge and self-discharge tests of the assembled LIB were successfully performed at a constant current of 10 μA, and the curve of the voltage change is shown in **Figure 3**. A maximum voltage of 3.6 V was recorded when the current was applied at intervals of 20 sec.

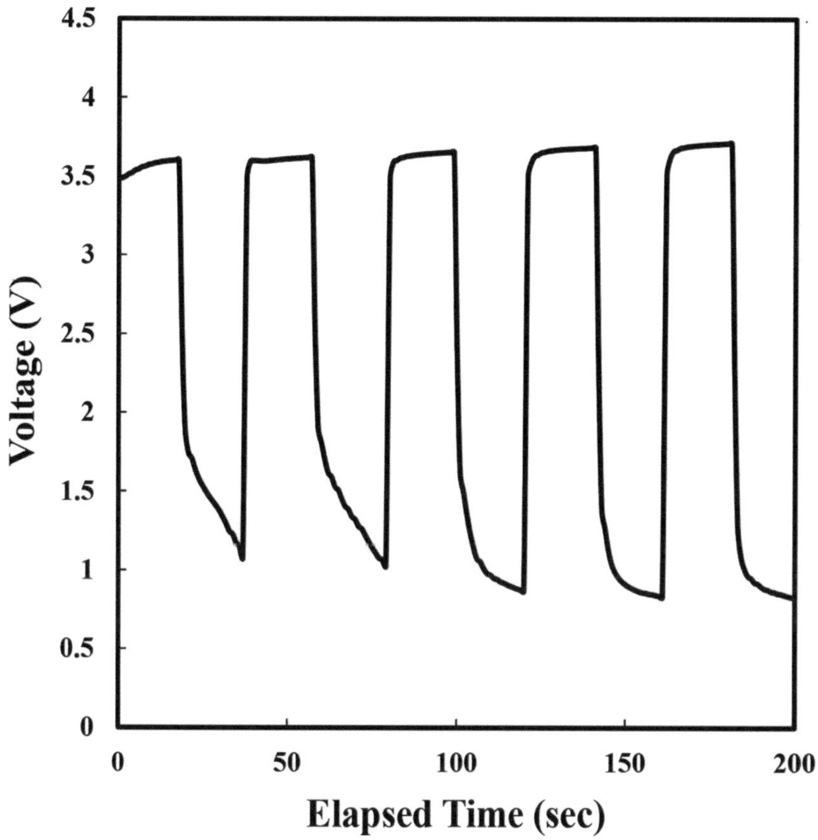

Figure 3. The repeated charge and discharge test of the battery. The charge at the constant current of 10 μA and spontaneous discharge was repeated at 20-s intervals.

When the battery was charged from an external source, the colorless battery drastically changed color to blue-gray, as shown in **Figure 4**. The color changes were repeatable and occurred simultaneously with the charge and discharge operations. The transmittance spectra of the charged and discharged battery are shown in **Figure 5**, and it is apparent that the colored battery after charging recovers repeatedly to a colorless one after spontaneous discharge. The specific coloration of the LTO electrode was confirmed by disassembling the colored battery.

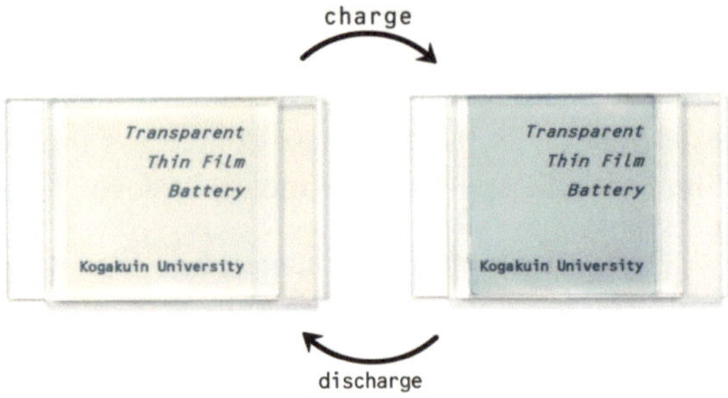

Figure 4. The colorless battery before charge and after discharge (left) and blue-gray battery after charge (right).

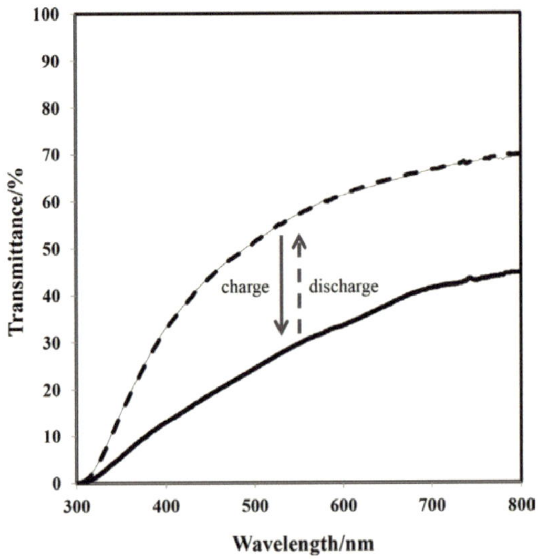

Figure 5. UV-vis spectra of the before/after charge and after discharge of the battery. The lines indicate the following:———, before charge; ▬▬▬▬, after charge; ▬ ▬ ▬, after discharge.

This unprecedented phenomenon suggests a two-step reaction based on the Ti^{4+}/Ti^{3+} redox coupling with the intercalation of Li^+ ions into the spinel-type LTO electrode. The electrochemical reaction can be described as follows [19]:

$$Li_4Ti_5O_{12} + 3Li^+ + 3e^- \rightarrow Li_7Ti_5O_{12}$$

The intercalation of Li^+ ions occurred in the vacant sites of the LTO spinel-skeleton through the electrolytic solution, and an equivalent amount of Li^+ ions were supplied from the LFP electrode by the charge operation. The coloration of the LTO to blue-gray in the process suggests that some of the Ti^{4+} sites were simultaneously reduced to Ti^{3+} ions by the electrons supplied from the power source. The Ti^{3+} ions could again be oxidized to Ti^{4+} ions along with the self-discharge of the battery, when the intercalated Li^+ ions returned to the LFP electrode through the electrolytic solution. Thus, the reversible LIB reaction could be visualized using the thin-film electrodes fabricated by this method. This monitoring system might be useful for clarifying the reaction mechanism of the novel LIB.

The rechargeable properties of this present LIB assembled with the newly fabricated ceramic thin films were examined by repeated charge and discharge testing. It is remarkable that the electrochromic reactions synchronous to the charge and discharge operations of the transparent thin-film battery can be clearly observed. Thus, the transparent thin-film device fabricated by the MPM may be applied to monitoring the redox reaction in electrochemical devices such as LIBs. The suitability of the MPM, as shown in this work, may also be useful for the production of transparent thin films of metal oxides and phosphates on large-surface-area materials.

4. Photovoltaic LIB

Titania has been used as a white pigment for paints, cosmetics, and foodstuffs [20]. When titania particles absorb UV radiation, they produce pairs of electrons and holes inside the particles. The surface states of the titania particle are important for photoreactivity because the photo-induced electrons and holes can be incorporated into redox reactions on the titania surface before spontaneous recombination [21–37].

Zhang et al. [38] reported that anatase particles immobilized on an anodic electrode constitute an excellent candidate for LIB active materials because of their high safety, good capacity retention, and low self-discharge. In fact, titania has negligible toxicity, excellent cyclability, and chemical stability in LIBs [39, 40]. Additionally, anatase nanotubes have also been characterized as active materials and have been shown to have electrochemical potentials of 1.5–1.7 V vs. Li/Li^+ [41]. We are, therefore, interested in the photovoltaic nature of LIBs, including transparent thin films of titania on their anodes because we previously fabricated both a transparent photocatalytic thin-film titania and the abovementioned LIB using the MPM. We attempted to fabricate a novel thin-film LIB that could be charged by light irradiation based on these results [42]. This novel, translucent, solar-chargeable LIB was fabricated using titania (anode) and $LiCoO_2$ (cathode) thin films prepared by MPM as the active materials on the above-mentioned conductive glass substrate. The precursor solutions containing the corresponding complexes capable of producing the anode and cathode active materials were easily prepared. Precursor films of TiO_2 and $LiCoO_2$ on the FTO pre-coated glass substrate were separately formed via a spin-coating method at ambient temperature using a two-step process, and they were pre-heated in a drying oven at 70 °C for 10 min. Then, the precursor films were heat treated in air for 30 min at 500 and 550 °C, respectively.

The X-ray diffraction peaks of the resulting thin films can be attributed to anatase and $LiCoO_2$. The optical transmittance of the assembled device was 50% at 700 nm, which is the longest wavelength in the visible region; hence, the device is translucent. A typical charge/discharge cyclic test was performed with a DC voltage source/monitor and was repeated 10 times at 20-sec intervals. The averaged potential at 2.34 V was observed by applying a constant current of 1.0 mA, then that at 2.01 V was detected after 20 sec during the sequential self-discharge process (**Figure 6**).

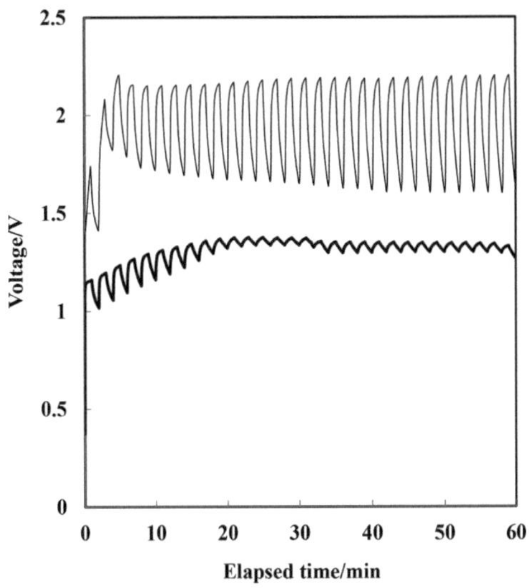

Figure 6. The charge/discharge cyclic test of the assembled battery. The lines indicate the following:———, a constant current of 0.2 mA; ▬▬▬▬, 1-sun irradiation.

Based on these plateau values, the potential difference between TiO_2 and $LiCoO_2$ can be theoretically estimated in the range of 2.3–2.0 V. Therefore, a device constructed of these active materials on a FTO pre-coated glass substrate could be operated as a typical LIB because the detected potentials in the charge/discharge cycles are in good agreement with the theoretical values. Therefore, the lithium intercalation/de-intercalation reactions in the active materials can be written as follows:

Anode: $TiO_2 + x\ Li^+ + x\ e^- \rightarrow Li_xTiO_2$

Cathode: $LiCoO_2 \rightarrow Li_{1-x}CoO_2 + xLi^+ + xe^-$

The repeated charge and self-discharge test of the LIBs performed in this study was also conducted under light irradiation and in the dark with no electrical supply 30 times at 60-sec intervals. The 1-sun irradiation was achieved using a solar simulator and monitoring with a DC voltage monitor. The irradiated area of the LIB was 4.0 cm^2. The voltage change during the test is also shown in **Figure 6**. The averaged voltages were 1.32 V during 1-sun irradiation and 1.29 V in the dark during the self-discharge process. Based on the calibration curve of the charging voltages over constant currents ranging from 0 to 1.0 mA, the detected value (1.38 V) can be theoretically reduced to the charging operation by applying a constant current of approximately 60 μA (**Figure 7**). Notably, the self-discharge voltage is almost identical to the charging voltage under light irradiation, indicating the effectiveness of the photovoltaic charge. The J-V curve of this LIB was measured under simulated sunlight AM 1.5 from a solar simulator with the radiant power of 1 -sun (**Figure 8**). It was found that the LIB generated http://dx.doi.org/ 10.5772/63491

photocurrent with short-circuit current; Isc = 7 × 10^{-5} A, open-circuit voltage; Voc = 0.97, fill factor; ff = 0.29 and a photo conversion efficiency of 0.02%. Thus, this LIB can also work as a solar cell.

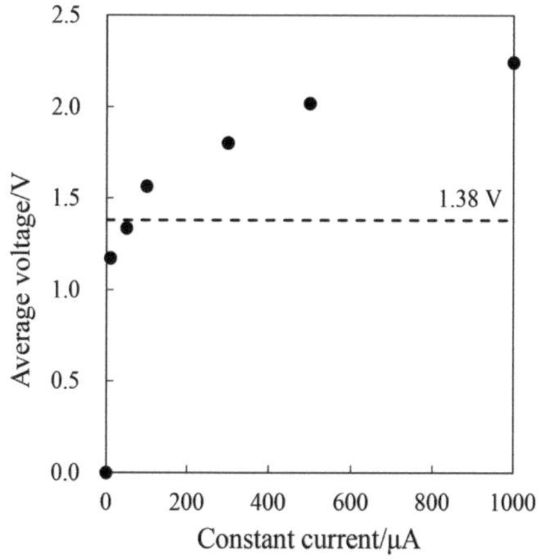

Figure 7. The averaged charging voltages over constant currents ranging from 0 to 1.0 mA.

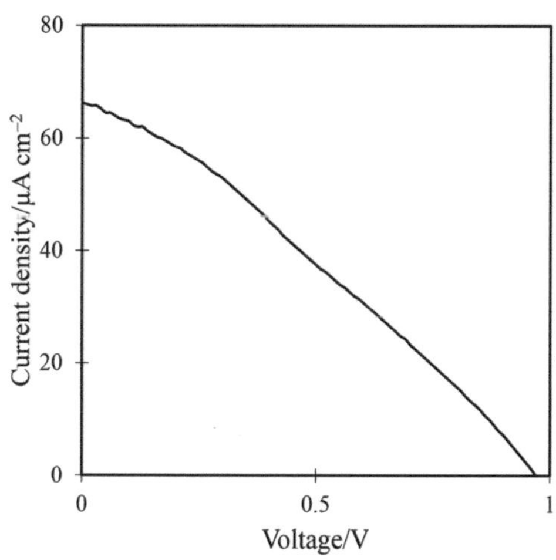

Figure 8. The photocurrent density-voltage characteristics (J-V curves) of LIB.

The current densities of the TiO_2 and $LiCoO_2$ on the FTO pre-coated glass substrate were measured at a bias of 0 V vs. Ag/AgCl to verify their conductive properties using a three-electrode system in an electrolytic solution. The results indicate that the TiO_2 thin film is an insulator in the dark and acts as an *n*-type semiconductor under 1-sun light irradiation based on the anodic photocurrent density observed (**Figure 9**), as we reported previously [43]. It can be assumed that the photovoltaic charging of the LIB requires the active materials on the anodic electrode to function as a semiconductor to produce electron-hole pairs and as lithium-ion conductors in their structures. Based on this assumption, there are numerous candidate active materials for photovoltaic LIBs. Investigating both the physical and chemical properties of active materials used in electrodes is thus required to successfully fabricate this type of innovative device.

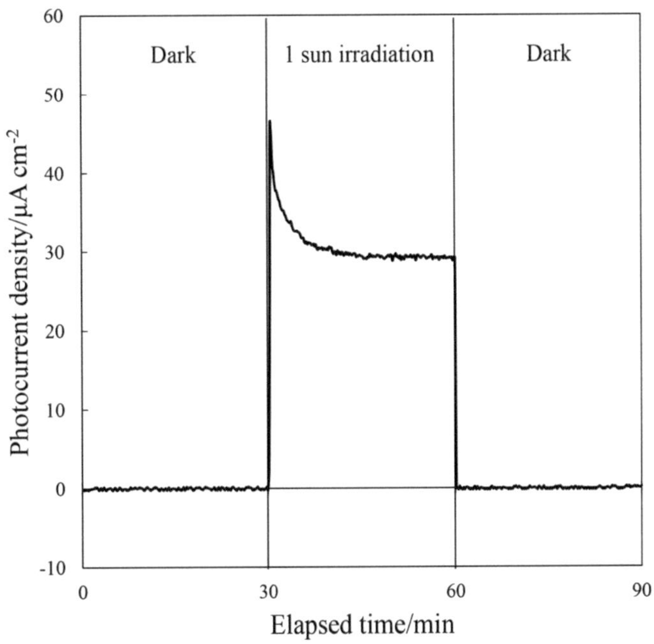

Figure 9. Photocurrent densities of anatase thin film fabricated on FTO glass. The photocurrent density of the sample electrode was measured under 1 sun light irradiation.

These results indicate that the translucent, two-in-one device can simultaneously generate and store electricity by irradiation with solar light and may be applicable for use as a multi-functional window, and facilitating the wider use of renewable energy.

5. Conclusion

The molecular precursor method is useful to obtain the active materials for the novel device which functioned as the lithium-ion battery, photovoltaic lithium-ion battery, and the solar cell. This is the original report on translucent, thin-film LIB that can be charged by solar light irradiation was fabricated using active materials on a conductive glass substrate.

The molecular precursor method was developed in our studies just 20 years ago. At that time, we reported the fabrication of Co_3O_4 and TiO_2 thin films. We subsequently reported the fabrication of thin films of various materials such as Cu_2O, SiO_2, ZnO, and apatite. This method is pertinent to coordination chemistry, materials science, nanoscience, and nanotechnology, and it has provided various thin films of high quality.

http://dx.doi.org/ 10.5772/63491

Author details

Hiroki Nagai and Mitsunobu Sato*

*Address all correspondence to: ft10302@ns.kogakuin.ac.jp

Department of Applied Physics, School of Advanced Engineering, Kogakuin University, Tokyo, Japan

References

[1] Wakihara M., Yamamoto O. Lithium ion batteries fundamentals and performance. Weinheim: WILEY–VCH Verlag GmbH; 1998.

[2] Murphy D. W., Carides J. N. Low voltage behaviour of lithium/metal dichalcogenide topochemical cells. J. Electrochem. Soc. 1979;126:349–351.

[3] Lazzari M., Scrosati B. A cyclable lithium organic electrolyte cell based on two intercalation electrodes. J. Electrochem. Soc. 1980;127:773–774.

[4] Nagaura T, Tazawa K. Lithium ion rechargeable battery. Prog. Batt. Solar Cells. 1990;9:20–29.

[5] Nagai H., et al. Synchronous electrochromism of lithium ion battery with chemically fabricated transparent thin films. Funct. Mater. Lett. 2013;6:1341001–1341008.

[6] Nagai H., Sato M. Heat treatment in molecular precursor method for fabricating metal oxide thin films. In: Frank C ed. Heat Treatment—Conventional and Novel Applications. Rijeka: InTech; 2012. pp. 103–128.

[7] Patouxa S., Wurm C., Morcrette M., Rousse G., Masquelier C. A comparative structural and electrochemical study of monoclinic $Li_3Fe_2(PO_4)_3$ and $Li_3V_2(PO_4)_3$. J. Power Sources. 2003;119:278–284.

[8] Yin S. C., Grondey H., Strobel P., Huang H., Nazar L. F. Charge ordering in lithium vanadium phosphates: electrode materials for lithium-ion batteries. J. Am. Chem. Soc. 2003;15:326–327.

[9] Masquelier C., Padhi A. K., Nanjundaswamy K. S., Goodenough J. B. New cathode materials for rechargeable lithium batteries: the 3-D framework structures $Li_3Fe_2(XO_4)_3$(X=P, As). J. Solid. State. Chem. 1998;135:228–234.

[10] Morgan D., Ceder G., Saidi M. Y., Barker J., Swoyer J., Huang H., Adamson G. Experimental and computational study of the structure and electrochemical properties of $Li_xM_2(PO_4)_3$ compounds with the monoclinic and rhombohedral structure. Chem. Mater. 2002;14:4684–4693.

[11] Colbow K. M., Dahn J. R., Haering R. R. Structure and electrochemistry of the spinel oxides $LiTi_2O_4$ and $Li_{4/3}Ti_{5/3}O_4$. J. Power Sources. 1989;26:397–402.

[12] Nishizawa M. Measurements of chemical diffusion coefficient of lithium ion in graphitized mesocarbon microbeads using a microelectrode, electrochemi. Solid State Lett. 1998;1:10–12.

[13] Ohzuku T., Iwakoshi Y., Sawai K. Formation of lithium-graphite intercalation compounds in nonaqueous electrolytes and their application as a negative electrode for a lithium ion (Shuttlecock) cell. J. Electrochem. Soc. 1993;140:2490–2497.

[14] Zaghib K., Armand M., Gauthier M. Electrochemistry of anodes in solid-state Li-ion polymer batteries. J. Electrochem. Soc. 1998;145:3135–3140.

[15] Ferg E. Spinel Anodes for lithium-ion batteries. J. Electrochem. Soc. 1994;141:147–150.

[16] Ohzuku T., Ueda A, Yamamoto A. Zero-strain insertion material of Li $[Li_{1/3}Ti_{5/3}]$ O_4 for rechargeable lithium cells. J. Electrochem. Soc. 1995:142:1431–1435.

[17] Kanamura K, Umegaki T., Naito H., Takehara Z., Yao T. et al., Structural and electrochemical characteristics of $Li_{4/3}Ti_{5/3}O_4$ as an anode material for rechargeable lithium batteries. J. Appl. Electrochem. 2001:31:73–78.

[18] Y. J. Hao et. al., Synthesis and characterization of spinel $Li_4Ti_5O_{12}$ anode material by oxalic acid-assisted sol–gel method. J. Power Sources 2006:158:1358–1364.

[19] Mosqueda H. A., Vazquez C., Bosch P., Pfeiffer H. Chemical sorption of carbon dioxide on lithium oxide. Chem. Mater. 2006;18:2307–2310.

[20] Hashimoto K., Irie H., Fujishima A. TiO_2 photocatalysis: A historical overview and future prospects. AAAPS. Buliten. 2007;17:12–28.

[21] Fujishima A., Honda K. Electrochemical photolysis of water at a semiconductor electrode. Nature. 1972;238:37–38.

[22] Serpone N., Pelizzetti E. Fundamental studies into primary events in photocatalysis employing CdS and TiO2 semiconductors: Photoluminescence, laser flash photolysis and pulse radiolysis. In: Homogeneous and Heterogeneous Photocatalysis. D. Reidel Publishing Company. Dordrecht; 1986. pp. 51–90.

[23] Serpone N., Pelizzetti E. Photocatalysis, Fundamentals and Applications. New York: John Wiley & Sons; 1989.

[24] Kamat P. V. Photochemistry on nonreactive and reactive (semi-conductor) surfaces. Chem. Rev. 1993;93:267–300.

[25] Ollis D. F. Photocatalytic purification and treatment of water and air. Al-Ekabi H ed. Amsterdam: Elsevier; 1993. 820 p.

[26] Heller A. Chemistry and applications of photocatalytic oxidation of thin organic films. Acc. Chem. Res. 1995;28:503–508.

[27] Hoffmann M. R., Martin S. T., Choi W., Bahnemann D. W. Environmental applications of semiconductor photocatalysis. Chem. Rev. 1995;95:69–96.

[28] Mills A., Lehunte S. An overview of semiconductor photocatalysis. J. Photochem. Photobiol. A: Chem. 1997;108:1–35.

[29] Peral J., Domènech X., Ollis D. F. Hetrogeneous photocatalysis for purification, decontamination and deodorization of air. J. Chem. Tech. Biotech. 1997;70:117–140.

[30] Fujishima A., Hashimoto K., Watanabe T. TiO_2 photocatalysis: fundamentals and applications. Tokyo: BKC, Inc; 1999. 184 p.

[31] Fujishima A., Rao T. N., Tryk D. A. Titanium dioxide photocatalysis. J. Photochem. Photobiol. C. 2000;1:1–21.

[32] Ollis D. F. Photocatalytic purification and remediation of contaminated air and water. C. R. Acad. Sci. Paris, Serie II C, Chim. 2000;3:405–411.

[33] Tryk D. A., Fujishima A., Honda K. Recent topics in photoelectrochemistry: achievements and future prospects. Electrochim. Acta. 2000;45:2363–2376.

[34] Fujishima A., Tryk D. A., Bard A. J., Stratmann M. Fundamentals of Photocatalysis. In: Licht S, ed. Encyclopedia of Electrochemistry. Vol. 6: Semiconductor Electrodes and Photoelectrochemistry. Weinheim; Wiley–VCH; 2002. 608 p.

[35] Hashimoto K., Irie H., Fujishima A. TiO_2 photocatalysis: a historical overview and future prospects. Japan. J. Appl. Phys. 2005;44:8269–8285.

[36] Fujishima A., Zhang X. Titanium dioxide photocatalysis: present situation and future approaches. C. R. Chimie. 2006;9:750–760.

[37] Fujishima A., Zhang X., Tryk D. A. Heterogeneous photocatalysis: from water photolysis to applications in environmental cleanup. Internat. J. Hydrogen Energy. 2007;32:2664–2672.

[38] Zhang H., et al. Electrochemical lithium storage of titanate and titania nanotubes and nanorods. J. Phys. Chem. C. 2007;111:6143–6148.

[39] Ortiz G. F., et al. Alternative Li-ion battery electrode based on self-organized titania nanotubes. Chem. Mater. 2009;21:63–67.

[40] Yang Z., et al. Nanostructures and lithium electrochemical reactivity of lithium titanites and titanium oxides: a review. J. Power Sources. 2009;192:588–598.

[41] Zakharova G. S., et al. Anatase nanotubes as an electrode material for lithium-ion batteries. J. Phys. Chem. C. 2012;116:8714–8720.

[42] Nagai H., Suzuki T., Takahashi Y., Sato M. Photovoltaic lithium ion battery fabricated by molecular precursor method. Funct. Mater. Lett. in print.

[43] Daniel L. S., Nagai H., Sato M. Absorption spectra and photocurrent densities of Ag nanoparticle/TiO_2 composite thin films with various amounts of Ag. J. Mater. Sci. 2013:48:7162–7170.

Chapter 7

Stress Analysis of Electrode Particles in Lithium-Ion Batteries

Yingjie Liu and Huiling Duan

Additional information is available at the end of the chapter

http://dx.doi.org/10.5772/62577

Abstract

This chapter reviews several theoretical models that are used to compute the stress fields inside the electrode particles of lithium-ion batteries during discharging/charging process and provides a guideline for researchers to choose the appropriate models. Due to the limitation of the existing models, a general electrochemo-mechanical framework is presented to model the concentration and stress fields of the electrode during the phase transformation. The interaction between stresses fields and phase transformation is addressed, which is a novel discovery in the research of lithium-ion batteries. The electrodes with different sizes and geometries are compared. The structural and electrochemical advantages of hollow core-shell structure particles are highlighted. The present work could help to accurate predict stress profile in electrode particles with different sizes, geometries, and charging operations and contributes to finding the optimal electrode. Therefore, this chapter is helpful for the material and structure design of electrodes of lithium-ion batteries.

Keywords: lithium-ion batteries, surface/interface stress, hydrostatic stress, phase transformation, core-shell electrode

1. Introduction

Lithium-ion batteries are the choices of diverse applications, such as electronics and electric cars because of their high capacity, high voltage, and long lifetime, and attract wide research interest in the community of chemistry, electro-chemistry, and mechanics [1–6]. During the process of charging/discharging, lithium ions insert into/extract from electrodes and induce high stresses [7]. The stress could cause the fracture of electrode when it exceeds the ultimate strength of the material [8–10]. This intercalation-induced fracture is indeed a key mechanism for lithium-ion battery's capacity fade.

To prevent this stress-induced electrode failure, accurate prediction of the stress fields is the first step. Therefore, multiple models are developed to address the effects of different factors on the stress field, e.g., hydrostatic stress [7], surface stress [11], charging operation [12, 13], material imperfection [14], plasticity [15, 16], heat generation [17, 18], particles' dissolution [19, 20]. Inserting the stress field into the fracture mechanics model, some novel models are proposed to investigate the electrode fracture and battery failure. For instance, Woodford et al. [14] calculates the stress intensity factor of the initial crack inside the electrode and the stability of crack growth. Several novel models are developed to predict the dynamic crack propagation [21–23]. The effects of some factors cannot be characterized by continuum models, which motivates some atomic scale research. Gao et al. developed the atomistic models to study the strong coupling of diffusion, stress, and solute concentration, and the surface locking instability during atomic intercalation into electrode [24, 25]. Suo et al. employed the first principle calculation to investigate the microscopic deformation and lithiation induced plasticity of silicon electrode [26, 27].

Phase transformation during the discharging/charging process has been widely reported in different electrode active materials [28, 29]. During the phase transformation, the electrode is divided into two phases by a phase interface. Because the equilibrium concentration of the two phases at the interface is different, an interface concentration discontinuity is observed. This phase interface is moving during the phase transformation, whose movement is characterized by an interface mass balance condition. Though this phase transformation can be tracked by the moving boundary models [30–33], the stress fields during the phase transformation are not well studied. Moreover, the effect of stress fields on phase transformation remains unknown. This chapter will systematically investigate the interactions between phase transformation and stress fields.

The second step to avoid the stress-induced electrode failure is to find an optimal electrode to lower the stress. The main research interests focus on the size and shape of the electrodes. Considerable efforts are put into studying the size of electrodes [25, 34]. The major conclusions are that nanoelectrode particles are not as easy to fail as micro ones, and the batteries with nanoelectrode particles have better cyclability. However, researchers obtain this conclusion by using the same models to analyze the behaviors of different-sized particles, which is incorrect. For example, the linear diffusion model [17] is only accurate for nanoelectrode particles while loses accuracy for large electrode particles. A comprehensive model that could provide accurate stress prediction for different sized electrode particles is needed. In terms of electrode shape, though researchers have tested the electrode particles of different shapes, e.g., spherical [34], core-shell [35, 36], nanowire [37], thin film [38, 39], layered plates [40], thin strip [41], and cylinder [42], it is still an open question what an optimal electrode is.

Hollow structure particle is a good candidate for the electrode of lithium-ion batteries, due to its unique structural properties, e.g., doubled surface area, core-shell structure, and large internal void [43]. Researchers have analyzed the mechanical properties of core-shell structure [44, 45], and emphasized its significant influence on the stress fields. Furthermore, the

http://dx.doi.org/10.5772/62577

advantages of hollow structure particles are highlighted by numerous experiments, when they are used as electrode particles. For example, hollow Si anode shows high initial discharging capacity and low-capacity degradation [46]. High capacity, good cyclability, and high rate capability are reported for hollow core-shell mesoporous TiO_2 spheres [48]. High coulombic efficiency, great rate performance, and excellent stability are observed in hollow Fe_2O_3 particles [49]. However, these works reveal few reasons for the optimal behaviors of hollow structure particles. Therefore, a comprehensive analysis is needed to investigate the properties of hollow structure electrode. In this chapter, a theoretical model on the hollow particles is introduced which considers the effects of hydrostatic stress, surface/interface stress, and phase transformation simultaneously.

In what follows, Section 2 briefly reviews the recently developed models on the stress fields of electrode particles. A guideline is proposed for researchers to choose appropriate models. Section 3 presents an electrochemo-mechanical framework to model the concentration and stress profile in the electrode of Lithium-ion batteries. The effects of hydrostatic stress, surface/interface stress, and phase transformation are fully coupled. Section 4 applies this framework to the hollow spherical electrode particle and calculates its concentrations and stress fields. In Section 5, the size and shape effects of electrode are analyzed to find the optimal electrode. Size effects induced by hydrostatic stress and surface/interface stresses are investigated through a cross-scale analysis, and shape effect is studied by varying the shell thickness. The structural and electrochemical advantages of hollow structure electrodes are investigated. Finally, some remarkable conclusions and discussions are provided in Section 6.

2. Review of existing models

To provide accurate predictions of the stress fields in the electrodes, multiple models are developed to address the effects of different factors on the stress field. This section briefly introduces some recently developed models and provides a guideline for researchers to choose appropriate models.

Zhang et al. [18] considers the effect of hydrostatic stress on the lithium flux, which provides accurate predictions on the stress fields. However, this effect is only important for micro electrode particles and can be ignored for nano ones. When the particle size is in nanoscale, Cheng et al. [11] reports the effect of surface stress, which is inversely proportional to the particle size. This surface effect is essential for nanoelectrode particles because it changes the stress state from traction free into compression and therefore prevents the growth of manufacture induced cracks. Therefore, the model with hydrostatic stress suits the research on micro electrode particles while the surface stress is a good choice for nano ones. Cheng et al. [12] and Lu et al. [50] investigate the influence of charging operations on stress fields, i.e., potentiostatic and galvanostatic operation. The galvanostatic process is the first stage of the charging process and occupies over 90% of the entire charging time and is commonly used in different models. However, there is some unused lithium ions at the end of galvanostatic process. To avoid the waste of the unused lithium, potentiostatic operation is needed. Therefore, potentiostatic operation is appropriate for the research on improving the battery efficiency.

The strain approximation is slightly different in modeling the cathode and anode. When modeling the cathode, since the deformation is small, the infinitesimal strain assumption is well accepted [14, 18, 35, 36]. However, during the charging/discharging process, the deformation of anode is large, especially when the anode material is silicon. Therefore, finite strain approximation is more reasonable in modeling the deformation of anode. The constitutive law is most important for the stress modeling, since it directly relates the stress state and deformation. Linear elastic assumption is usually the choice, since it is the simplest and the deformation of electrode particles usually stays in the elastic range. However, when the deformation is large enough, the stress can exceed the yielding criterion and the deformation may reach plastic scope. Therefore, linear elastic models are used for small deformation modeling, while plastic models are usually employed in modeling large deformation of electrode [15, 16].

Manufacture-induced initial flaws or cracks are important reasons for the failure of electrodes and therefore need to be carefully addressed. Stress concentration appears at these imperfections, and the growth of the imperfections could lead to the fracture of the electrodes. Woodford et al. proposes a fracture mechanics model to predict the stress fields at the initial crack [14]. However, this model can only determine whether the initial crack would grow. To model the dynamic crack propagation progress, multiple fracture models are developed [21–23]. In addition to cracks, one can investigate other failure mechanisms of the electrodes, e.g., delamination [13, 50], by inserting the stress fields into the corresponding models.

Phase transformation is experimentally observed in the electrode active material during discharging/charging process and could significantly affect the stress fields [30–32]. Multiple models have been developed to address the phase transformation-induced discontinuities in concentration and stress fields. However, the inverse effect of the stress fields on the phase transformation has not been well studied, until the recent work of current authors [47]. Liu et al. [47] proposes a fully coupled system which investigates the interactions between phase transformation and stress fields, which will be discussed in this chapter.

In the following, this chapter proposes an electrochemo-mechanical framework to model the concentration and stress fields of electrode particles. The interactions between stress fields and phase transformations are characterized. The effect of hydrostatic stress, surface stress, and interface stress are fully addressed.

3. Electrochemo-mechanical framework

There are three different stages in the whole discharging process, which are schematically illustrated in **Figure 1**. In the first stage, there is α phase only (c.f. **Figure 1a**). The lithium ion inserts into the electrode particles from the outer surface, and its concentration increases with time. This stage finishes when the concentration at the outer surface reaches the equilibrium value of α phase (c_{eq}^{α}). In the second stage, the α phase electrode material gradually transforms into β phase material, and the two phases are separated by a sharp phase interface $S_{\alpha\beta}$ (c.f. **Figure 1b**). The concentrations of the two phases equal their equilibrium values at the interface, i.e., $c^{\alpha}\big|_{S_{\alpha\beta}}=c^{\alpha}_{eq}$, and $c^{\beta}\big|_{S_{\alpha\beta}}=c^{\beta}_{eq}$. Because $c^{\alpha}_{eq}\neq c^{\beta}_{eq}$, a concentration discontinuity appears at the interface. The flux is also discontinuous at the phase interface, which drives the interface to move inward. The phase transformation finishes when the material of the whole electrode particle changes into β phase. At this time, the concentration of lithium ions at the inner surface equals c^{β}_{eq}. The third stage is similar to the first one, in which there is β phase only in the electrode particle and the concentration increases with time (c.f. **Figure 1c**). Please note that the charging process can also be divided into three stages: (1) β phase only, (2) biphase, (3) α phase only, and therefore is the opposite process of discharging.

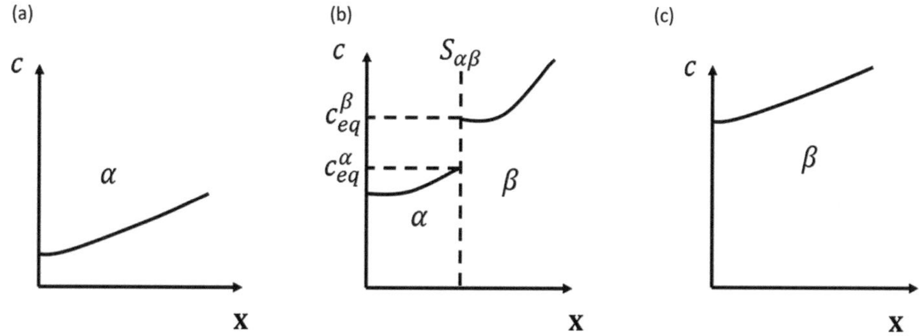

Figure 1. Schematic illustration of lithium concentration inside the electrode particle during discharging process: (a) first stage, there is α phase only and the concentration is continuous; (b) second stage, α and β phases coexist and are separated by a sharp interface $S_{\alpha\beta}$ at which the concentration is discontinuous; (c) third stage, there is β phase only. Please note that the concentration is discontinuous at second stage. Here x and c denote the position and lithium ions' concentration, c^{α}_{eq} and c^{β}_{eq} represent the equilibrium values of the two phases.

3.1. Mechanical equations

As illustrated in Figure 1b, a phase interface $\mathbf{S}_{\alpha\beta}$ appears in the second stage of the discharging process. The electrode particle is separated into $\boldsymbol{\alpha}$ and $\boldsymbol{\beta}$ phases by $\mathbf{S}_{\alpha\beta}$ (c.f. **Figure 2**). Inserting lithium ions' into and extracting them from the electrode particle will cause its non-uniform distribution and then induce stress fields. Therefore, the concentration of lithium ions is considered in the constitutive law to characterize this mechanics-electrochemical coupling problem. For both $\boldsymbol{\alpha}$ and $\boldsymbol{\beta}$ phases, the constitutive law can be written as [50]

$$\boldsymbol{\varepsilon}^{p}=\frac{1}{E^{p}}\left[\left(1+v^{p}\right)\boldsymbol{\sigma}^{p}-v^{p}\Theta^{p}\boldsymbol{I}\right]+\frac{\tilde{c}^{p}\Omega^{p}}{3}\boldsymbol{I},\ \left(p=\alpha,\beta\right) \tag{1}$$

where $\boldsymbol{\varepsilon}^{p}$ and $\boldsymbol{\sigma}^{p}$ denote strain and stress tensors of phase p, and $\Theta^{p}=\sum_k\sigma^{p}_{kk}$. E^{p}, v^{p} and Ω^{p} are the Young's modulus, Poisson's ratio, and partial Molar volume of phase p, respectively. is the change of lithium ion's concentration $c^{p}(x, t)$ from the initial value $c^{p}_{0}(x)$ in the electrode particles. Here, and in the following, the superscript p = α, β denote the α and β phases, respectively.

Classical elasticity theory yields the equilibrium equation,

$$\nabla \cdot \boldsymbol{\sigma}^{p} = 0 \tag{2}$$

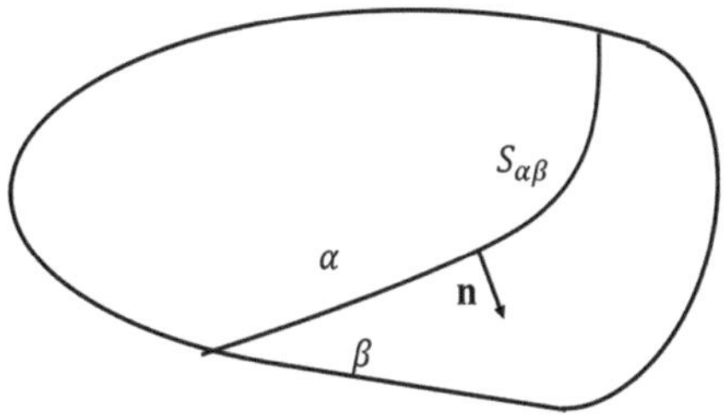

Figure 2. Schematic illustration of electrode particle: an interface $S_{\alpha\beta}$ separates the electrode particle into α and β phases.

and the infinitesimal strain geometric equation

$$\boldsymbol{\varepsilon}^{p} = \frac{1}{2}\left(\nabla \boldsymbol{u}^{p} + \boldsymbol{u}^{p}\nabla\right) \tag{3}$$

where $\mathbf{u}^p$ is the displacement vector of phase p.

Combination of Eqs. (1)–(3) yields the governing equation of each phase. In addition to the equations of two bulk phases, the deformation of interface needs to be characterized. The equilibrium equation of the phase interface $S_{\alpha\beta}$ is [51]

$$[\boldsymbol{\sigma}] \cdot \boldsymbol{n} = -\nabla_{s} \cdot \boldsymbol{\tau} \tag{4}$$

where $[\boldsymbol{\sigma}] = \boldsymbol{\sigma}^{\beta}\big|_{S_{\alpha\beta}} - \boldsymbol{\sigma}^{\alpha}\big|_{S_{\alpha\beta}}$, $\boldsymbol{\sigma}^{\alpha}$ and $\boldsymbol{\sigma}^{\beta}$ are the stress tensors of two phases, respectively. $\nabla_s \cdot \boldsymbol{\tau}$ denotes the interface divergence of interface stress $\boldsymbol{\tau}$ at the phase interface. Please note that stress discontinuity was only considered at the interface of two materials before [51–54], and the present chapter introduces the stress discontinuity at the phase interface of electrode in lithium-ion battery.

The interface constitutive law yields the relation between interface stress tensor $\boldsymbol{\tau}$ and interface strain tensor $\boldsymbol{\varepsilon}^s$

$$\boldsymbol{\tau} = \tau^{0}\mathbf{1} + \boldsymbol{C}^{s} : \boldsymbol{\varepsilon}^{s} \tag{5}$$

http://dx.doi.org/10.5772/62577

where τ^0 is the strain-independent interface stress, **1** is a 2D second-order unit vector, and $\mathbf{C}^s$ is the interface stiffness tensor. Here and in the following, the super-subscript "s" denotes the interface/surface quantities.

The two phases are fully bonded at the interface, which yields the no jump condition

$$\boldsymbol{u}^{\alpha}\Big|_{S_{\alpha\beta}} = \boldsymbol{u}^{\beta}\Big|_{S_{\alpha\beta}} \tag{6}$$

where $\mathbf{u}^\alpha$ and $\mathbf{u}^\beta$ are the displacement vectors of two phases, respectively. Solving Eqs. (1)–(3) for α and β bulk phases, and Eqs. (4)–(6) for the interface $S_{\alpha\beta}$, one can compute the stress profile of the biphase electrode particle.

3.2. Electrochemical equations

Lithium ions extract from/insert into electrode particles during the charging/discharging process. This process is usually modeled as the diffusion of lithium ions, driven by the gradient of chemical potential. The velocity ($\mathbf{V}^p$) of lithium ions in phase p can be written as follows [18]:

$$\boldsymbol{V}^p = -M^p \nabla \phi^p, \quad (p = \alpha, \beta) \tag{7}$$

where $\mathbf{M}^p$ and ϕ^p are the mobility and electrochemical potential. The species flux ($\mathbf{J}^p$) is

$$\boldsymbol{J}^p = c^p \boldsymbol{V}^p = -c^p M^p \nabla \phi^p \tag{8}$$

where the electrochemical potential ϕ^p in an ideal solid solution is [18],

$$\phi = \phi_0^p + RTlnX - \Omega^p \sigma_h^p \tag{9}$$

where ϕ_0^p is a constant, R is gas constant, T is absolute temperature, X is molar fraction of lithium ion, and σ_h^p is hydrostatic stress,

$$\sigma_h^p = \frac{1}{3}\sigma_{kk}^p \tag{10}$$

Substituting Eq. (9) into Eq. (8), one obtains the species flux,

$$\boldsymbol{J}^p = -M^p c^p \left(RT\frac{1}{X}\nabla X - \Omega^p \nabla \sigma_h^p \right) = -D^p \left(\nabla c^p - \frac{\Omega^p c^p}{RT}\nabla \sigma_h^p \right) \tag{11}$$

where $D^p = M^pRT$ is diffusivity. Please note that the effect of hydrostatic stress on the species flux was neglected by previous work [11, 12, 41, 42], and is first proposed in [18]. The equation of substance conservation is written as

$$\frac{\partial c^p}{\partial t} + \nabla \cdot \boldsymbol{J}^p = 0 \tag{12}$$

Inserting Eq. (**11**) into Eq. (**12**), the governing equation of Li ions' concentration is obtained,

$$\frac{\partial c^p}{\partial t} - D^p\left(\nabla c^p - \frac{\Omega^p c^p}{RT}\nabla \sigma_h^p\right) = 0 \tag{13}$$

The boundary conditions are determined by the charging operation. Under galvanostatic operation,

$$\boldsymbol{J}^p \cdot \boldsymbol{n} = \frac{i_n}{F}(\text{active}); \quad \boldsymbol{J}^p \cdot \boldsymbol{n} = 0(\text{inactive}) \tag{14}$$

where **n** is the out normal of the boundaries. F is Faraday constant and $\mathbf{i}_n$ is currency density, respectively. When phase transformation begins, a phase interface appears between $\boldsymbol{\alpha}$ and $\boldsymbol{\beta}$ phases, as illustrated by Figure **2**. At the phase interface, the concentration of each phase equals its equilibrium value, i.e.,

$$c^\alpha\Big|_{S_{\alpha\beta}} = c_{eq}^\alpha; \quad c^\beta\Big|_{S_{\alpha\beta}} = c_{eq}^\beta \tag{15}$$

The movement of the phase interface is assumed to be under the control of diffusion process in the adjacent phases, and the interface position is tracked by a jump material balance [56],

$$\left(c_{eq}^\alpha - c_{eq}^\beta\right)\boldsymbol{v} \cdot \boldsymbol{n}^\alpha = \left(\boldsymbol{J}^\alpha - \boldsymbol{J}^\beta\right) \cdot \boldsymbol{n}^\alpha \tag{16}$$

where **v** is moving velocity of the phase interface, while $\mathbf{n}^\alpha$ is the normal direction from phase α pointing to phase β. Therefore, $\mathbf{v} \cdot \mathbf{n}^\alpha = \mathbf{v}_n$ is normal velocity which is negative in the discharging process and positive in the charging process. Noted that Eq. (16) is automatically satisfied when there is only one phase in the particle. Substitution of Eq. (11) into (16) yields the governing equation of the interface,

http://dx.doi.org/10.5772/62577

$$\left(c_{eq}^{\alpha}-c_{eq}^{\beta}\right)v_n = D^{\beta}\left(\frac{\partial c^{\beta}}{\partial n}-\frac{\Omega^{\beta}c^{\beta}}{RT}\frac{\partial\sigma_h^{\beta}}{\partial n}\right)-D^{\alpha}\left(\frac{\partial c^{\alpha}}{\partial n}-\frac{\Omega^{\alpha}c^{\alpha}}{RT}\frac{\partial\sigma_h^{\alpha}}{\partial n}\right) \tag{17}$$

where $\partial/\partial n$ denotes the gradient along the normal direction of interface. Please note that the effect of σ_h on the movement of phase interface has not been proposed in the literature [40, 41].

The electrode particle is initially at the stress free state, which implies that the initial concentration is uniformly distributed, i.e.,

$$c^{p}\left(t=0,x\right)=c_0 \tag{18}$$

A general electrochemo-mechanical framework is developed for an electrode particle of arbitrary geometry in lithium-ion batteries. The discharging/charging process can be regarded as a quasi-static process. In each time step, by inserting the concentration of Li ions into Eqs. (1)–(6), one can compute the stress field. Substituting the stress field into Eqs. (13)–(17), one can obtain the concentration field.

4. Stress field in hollow spherical electrode particle

Section 3 provides a electrochemo-mechanical framework to compute the concentration and stress profile for an electrode particle of arbitrary geometry in lithium ion batteries. This section will perform analysis on the concentration and stress fields of a specific electrode particle. Hollow structures have great potential as electrode of lithium-ion batteries, due to their better cyclability, higher capability, and lower capacity degradation [46, 48, 49]. Therefore, in the following, we will analyze the concentration and stress fields of the hollow electrode particles.

Figures 3 illustrate the structure of hollow spherical particles, r_1 and r_2 are the inner and outer radii of the hollow particles, respectively. $\zeta = r_1/r_2$ denotes the ratio of inner and outer radii. As emphasized in last section, α and β phases coexist in the electrode particle in the second stage of discharging process (c.f. **Figure 3a**). For this biphase particle, the inner layer is α phase and the outer one is β phase, which are separated by a sharp interface. Please note that the biphase electrode particle reduces to a single-phase electrode particle, when the phase interface overlaps with the outer/inner surfaces (c.f. **Figure 3b and 3c**). In what follows, the proposed electrochemo-mechanical framework in section 3 is applied on this specific hollow spherical particle electrodes, and the concentration and stress fields are analyzed.

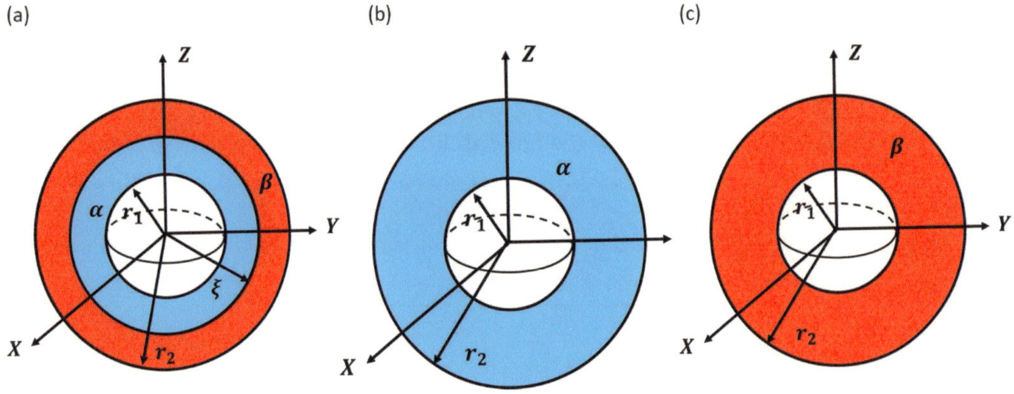

Figure 3. Three stages of the hollow $LiCoO_2$ particles in discharging process: (a) α and β phases coexist; (b) α phase only; (c) β phase only (reproduction from [47]).

4.1. Electrochemo-mechanical framework of hollow particle

As an axisymmetric problem, the displacement vector only has radial component (u) and the stress and strain tensors only contain two independent components σ_r , σ_θ and ε_r , ε_θ (note $\sigma_\phi = \sigma_\theta$, $\varepsilon_\phi = \varepsilon_\theta$). Under the spherical coordinate, Eqs. (1)–(3) become

Constitutive Law:

$$\begin{cases} \varepsilon_r^p = \dfrac{1}{E^p}\left(\sigma_r^p - 2\nu^p \sigma_\theta^p\right) + \dfrac{\Omega^p}{3}\tilde{c}^p \\ \varepsilon_\theta^p = \dfrac{1}{E^p}\left(\sigma_\theta^p - 2\nu^p\left(\sigma_r^p + \sigma_\theta^p\right)\right) + \dfrac{\Omega^p}{3}\tilde{c}^p \end{cases} \tag{19}$$

Equilibrium equation:

$$\frac{d\sigma_r^p}{dr} + \frac{2}{r}\left(\sigma_r^p - \sigma_\theta^p\right) = 0 \tag{20}$$

Geometric equation:

$$\varepsilon_r^p = \frac{du^p}{dr}; \quad \varepsilon_\theta^p = \frac{u^p}{r} \tag{21}$$

Combining Eq. (19)–(21), one could solve for the stress fields, i.e.,

$$\begin{cases} \sigma_r^p = E^p\left[\dfrac{-2}{1-\nu^p}\dfrac{\Omega^p}{9}c_{av}^p + \dfrac{A^p}{1-2\nu^p} - \dfrac{2B^p}{r^3\left(1+\nu^p\right)}\right] \\ \sigma_\theta^p = E^p\left[\dfrac{1}{1-\nu^p}\dfrac{\Omega^p}{9}c_{av}^p - \dfrac{1}{1-\nu^p}\dfrac{\Omega^p}{3}\bar{c} + \dfrac{A^p}{1-2\nu^p} + \dfrac{B^p}{r^3\left(1+\nu^p\right)}\right] \end{cases} \tag{22}$$

where $c_{av}^p = \frac{3}{r^3}\int_{r_1^p}^{r} p\, r^2 dr$. Please note that $r_1^\alpha = r_1$ and $r_1^\beta = \xi$, and ξ is the position of interface. A^p, B^p are constants that need to be determined via the boundary conditions.

When α and β phase coexist, there is a phase interface appearing at r = ξ. Figure 3a provides a schematic illustration of the phase distribution at this stage.

For the hollow spherical electrode particle, the interface equilibrium Eq. (4) is given as

$$\left(\sigma_r^\beta - \sigma_r^\alpha\right)\Big|_{r=\xi} = \frac{2\tau_\theta}{\xi} \tag{23}$$

For the isotropic interface, the interface constitutive law Eq. (5) is

$$\tau_\theta = \tau^0 + K^s \varepsilon_\theta^p \tag{24}$$

where K^s is interface modulus ($\varepsilon_\phi^p = \varepsilon_\theta^p$). The displacement continuity Eq. (6) reduces to,

$$u^\alpha\Big|_{r=\xi} = u^\beta\Big|_{r=\xi} \tag{25}$$

The surface stress is effected via the boundary condition [55],

$$\sigma_r^p\Big|_{r=r_1} = \frac{2\tau_\theta}{r_1};\quad \sigma_r^p\Big|_{r=r_2} = \frac{2\tau_\theta}{r_2} \tag{26}$$

It is noted that this boundary condition can be regarded as the special case of Eq. (4), when the interface is located at the inner and outer surfaces.

The constants A^α, B^α, A^β, and B^β are determined by solving Eqs. (23)–(26). Inserting them into Eq. (22), one can calculate the stress. When there is only α or β phase in the electrode particle, $A^\alpha = A^\beta$, $B^\alpha = B^\beta$, and Eq. (25) is automatically satisfied. In this case, phase interface ξ can be regarded as r_2 for electrode with α phase only, while as r_1 for β phase only (illustrated by Figures 3b and 3c). Moreover, the interface condition Eq. (23) is automatically satisfied for the single phase electrode particle because it can reduce to the boundary condition Eq. (26). Therefore, one can obtain the solution of the single phase particle by setting ξ equals r_2 (α phase only) or r_1 (β phase only) in the solution of biphase particle.

The stresses inside the electrode particles are induced by the non-uniform distribution of concentration, which is studied by an electrochemical model. In the first and third stages of the discharging process, there is only α or β phase in the electrode particle (c.f. Figures 3b and 3c). For this single axisymmetric electrode, the governing equation of phase p Eq. (12) reduces to

$$\frac{\partial c^p}{\partial t}=\frac{D^p}{r^2}\frac{\partial}{\partial r}\left(r^2\frac{\partial c^p}{\partial r}-\frac{c^p\Omega^p}{RT}r^2\frac{\partial \sigma_h^p}{\partial r}\right) \tag{27}$$

The gradient of the hydrostatic stress can be rewritten as

$$\frac{\partial \sigma_h^p}{\partial r}=-\frac{2E^p}{1-\nu^p}\frac{\Omega^p}{9}\frac{\partial c^p}{\partial r} \tag{28}$$

Substituting Eq. (28) into Eq. (27), the final governing equation is obtained,

$$\frac{\partial c^p}{\partial t}=\frac{D^p}{r^2}\frac{\partial}{\partial r}\left(\left(1+\theta^p c^p\right)r^2\frac{\partial c^p}{\partial r}\right) \tag{29}$$

where $\theta^p = 2E^p(\Omega^p)^2/[9(1-\nu^p)RT]$. Rewrite the dimensionless parameter $\lambda^p = \theta^p c^p = \theta^p c_{max} \cdot c^p/c_{max}$. Considering $c^p/c_{max} \in (0, 1)$, one can expect that the effect of hydrostatic stress depends on the value of λ^p. Define the effective diffusivity $D_{\text{eff}}^p = D^p(1+\theta^p c^p)$, and rewrite Eq. (29) as

$$\frac{\partial c^p}{\partial t}=\frac{1}{r^2}\frac{\partial}{\partial r}\left(r^2 D_{\text{eff}}^p\frac{\partial c^p}{\partial r}\right) \tag{30}$$

One can observe that the governing equation could reduce to a similar form to the classical diffusion equation. The effects of hydrostatic stress are effected via a concentration dependent diffusivity. The lithium ion flux is inserted into and extracted from the electrode through the outer surface. So the outer surface is active, while inner surface is inactive. Therefore, the specific form of boundary condition Eq. (14) is

http://dx.doi.org/10.5772/62577

$$-D_{\mathrm{eff}}^{p}\frac{\partial c^{p}}{\partial r}\bigg|_{r=r_1}=0;\quad -D_{\mathrm{eff}}^{p}\frac{\partial c^{p}}{\partial r}\bigg|_{r=r_2}=\frac{i_n}{F} \tag{31}$$

Eq. (31) implies that the concentration gradient at the outer surface is inversely proportional to the factor $1 + \theta^p c^p$.

When the concentration at the outer surface reaches the equilibrium value of the α phase, phase transformation begins. Figure 3a illustrates the biphase hollow electrode particle. Combining the diffusion process of two phases, one can obtain the governing equations from Eq. (29),

$$\begin{cases} \dfrac{\partial c^{\alpha}}{\partial t}=\dfrac{1}{r^2}\dfrac{\partial}{\partial r}\left(r^2 D_{\mathrm{eff}}^{\alpha}\dfrac{\partial c^{\alpha}}{\partial r}\right) & r_1\le r\le \xi \\ \dfrac{\partial c^{\beta}}{\partial t}=\dfrac{D^{\beta}}{r^2}\dfrac{\partial}{\partial r}\left(r^2 D_{\mathrm{eff}}^{\beta}\dfrac{\partial c^{\beta}}{\partial r}\right) & \xi\le r\le r_2 \end{cases} \tag{32}$$

For the hollow spherical particles, the governing equation of the moving interface Eq. (17) reduces to

$$\left(c_{eq}^{\alpha}-c_{eq}^{\beta}\right)\frac{d\xi}{dt}=D_{\mathrm{eff}}^{\beta}\frac{\partial c^{\beta}}{\partial r}-D_{\mathrm{eff}}^{\alpha}\frac{\partial c^{\alpha}}{\partial r} \tag{33}$$

where $d\xi/dt$ is the moving velocity of the phase interface. Note that the contributions of θ^α and θ^β to the movement of phase interface have been ignored in the previous work. One can regard the current version of moving interface equation as the same as the one in [33] but replacing the constant diffusivity D^p with an effective diffusivity D_{eff}^{p}.

The concentrations at two sides of the interface are equal to their own equilibrium values, i.e.,

$$c^{\alpha}\big|_{r=\xi}=c_{eq}^{\alpha};\quad c^{\beta}\big|_{r=\xi}=c_{eq}^{\beta} \tag{34}$$

At this stage, the α phase satisfies the boundary condition at the inner surface, while β phase satisfies the boundary condition at the outer one, i.e.,

$$-D_{\mathrm{eff}}^{\alpha}\frac{\partial c^{\alpha}}{\partial r}\bigg|_{r=r_1}=0;\quad -D_{\mathrm{eff}}^{\beta}\frac{\partial c^{\beta}}{\partial r}\bigg|_{r=r_2}=\frac{i_n}{F} \tag{35}$$

During the phase transformation process, the α phase gradually transforms into β phase, and the phase interface moves inward. When the phase transformation finishes, the phase interface reaches the inner surface ($\xi = r_1$), and only β phase is left in the electrode particle (c.f. Figure 3c).

Initially, the lithium ions are uniformly distributed inside the electrode particle, i.e.,

$$c^{p}\left(r,t=0\right)=c^{0},\quad r\in\left[r_1,r_2\right] \tag{36}$$

The beauty of the formulation for the hollow spherical particle is that the governing equations Eqs. (29), (32), and (33) are decoupled from the stress field explicitly. This implies that one can directly compute the concentration field by using Eqs. (29)–(36) instead of solving a fully coupled system, and then calculate the stress field using Eqs. (22)–(26) at each step.

4.2. Concentration and stress fields

Section 4.1 provides the model to compute concentration and stress fields. The effect of hydrostatic stress is coupled into governing equation Eq. (32), and the surface/interface stress are effected via Eq. (23) and (26). To reveal the interactions between stress profile and phase transformation, a hollow spherical electrode particle is analyzed, for example. The inner and outer radii of electrode particle are 10 and 20 μm, and the particle is discharged under 10 A/m^2 currency density. Here and in the following, $LiCoO_2$ is used as the electrode material unless otherwise specified because it is the mostly reported electrode material with phase transformation phenomenon and most widely used in industry. The material properties of $LiCoO_2$ are provide in [32].

4.2.1. Concentration field

The concentration field of the electrode particle at the three stages are calculated (c.f. Figure 4). At the first stage, there is α phase only. **Figure 4a** illustrates the lithium ions' concentration at different time in this stage. Observations show that the concentration increases monotonically with r from the inner surface to the outer one. Because of the inward flux at the outer surface, the concentration keeps increasing with time. However, the concentration gradient decreases with time. This is consistent with Eq. (*31*) which implies that the concentration gradient is inversely proportional to the factor $1+\theta c$. Therefore, the concentrations at the inner and outer surfaces are significantly different at 50 seconds, but close to each other at 500 seconds. The first stage finishes at 2106.5 seconds, when the concentration at the outer surface reaches c_{eq}^{α}, and the phase transformation begins.

When the phase transformation happens, a phase interface appears between the $\boldsymbol{\alpha}$ and $\boldsymbol{\beta}$ phases. Driven by the interface flux jump, the phase interface gradually moves towards the inner surface. **Figure 4b** illustrates the concentration distribution when the phase interface is at different positions. The concentrations of the two phases at the interface equal their own equilibrium values, i.e., c_{eq}^{α} and c_{eq}^{β}. Because $c_{eq}^{\alpha}=0.75\,c_{max}$ and $c_{eq}^{\beta}=0.97c_{max}$, a sharp concentration jump is observed at the phase interface. Similar to the first stage, the concentration increases

http://dx.doi.org/10.5772/62577

with time during the phase transformation. However, under the effect of concentration discontinuity, the concentration gradient also increases with time. This is because the concentration increasing at the outer surface is slower than the movement of phase interface. When the phase transformation finishes, the concentration is c_{eq}^{β} at the inner surface, while close to c_{max} at the outer surface.

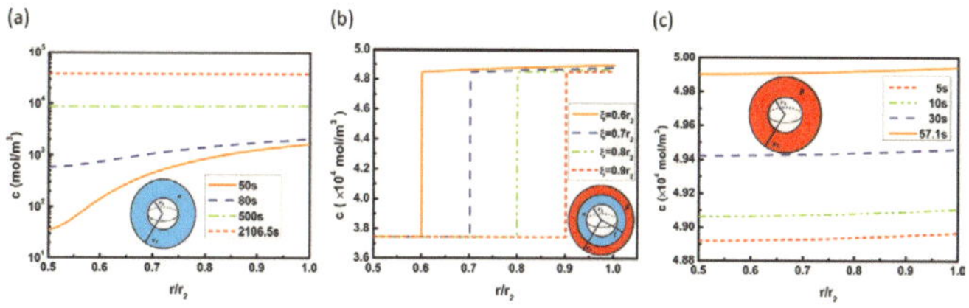

Figure 4. Concentration distribution of the electrode particle: (a) with α phase only; (b) with both α and β phases; (c) with β phase only (some data are reproduced from [47]).

To address the interactions between phase transformation and stress field, one can first investigate the effect of hydrostatic stress on the phase transformation. Please recall that the governing equation of the phase interface is modified to include the gradient of hydrostatic stress. Table 1 lists the time it needs for the phase interface to arrive the positions illustrated in Figure 4b. Compared with the conventional equations without considering the effect of hydrostatic stress, the moving speed of phase interface is much faster, which implies that the time during the phase transformation stage is overestimated by the previous models [33]. This is because the factor $(1 + \boldsymbol{\theta} \mathbf{c})$ significantly increases the flux jump at the interface, i.e., the driven force of interface movement. In other words, one can conclude that the effect of hydrostatic stress can significantly accelerate the phase transformation.

Position	Without hydrostatic stress	With hydrostatic stress
$0.9\ r_2$	23,500 s	171 s
$0.8\ r_2$	44,950 s	328 s
$0.7r_2$	61,210 s	452 s
$0.6r_2$	73,700 s	550 s

Table 1. Time needed for the movement of phase interface.

When the phase interface arrives at the inner surface, the phase transformation ends and the third stage begins. There is β phase only in the electrode at this stage. The concentration distribution at different time is illustrated in **Figure 4c**. Because the concentration at outer surface has been close to c_{max} at the end of second stage, the third stage only takes about 57 seconds. Similar to the first stage, concentration increases with time while its gradient decreases with time. In the end, the concentration increases with r, and is close to c_{max}

everywhere. Please note this is only the end of galvanostatic process, which is followed by potentiostatic operation to fully discharge the battery.

4.2.2. Stress profile

Lithium ion is non-uniformly distributed in the electrode and induces stresses. The stress field of a single phase electrode particle has been investigated by many researchers [18, 21, 35, 36], while the stress field of the biphase one is still unknown. Eq. (22) implies that the stress fields are explicitly related to the concentration field. One can expect the interface concentration discontinuity to have a significant influence on the stress field, which is illustrated in **Figure 5**. The constant surface stress $\tau^0 = 1J/m^2$ and surface modulus $K^s = 5N/m$, unless otherwise specified [11]. The radial stress of the single phase electrode particle monotonically increases/decreases in the discharging/charging process [14, 18]. But the radial stress of this biphase particle increases from the inner surface to the phase interface and decreases from the interface to the outer surface. The stresses predicted from the present model are compared with those from the one without considering surface/interface effect. Without considering the surface effect, radial stress is zero at the surfaces because Eq. (26) reduces to the traction free boundary condition. Under the effect of surface stress, the radial stress is positive at the inner surface, and negative at the outer one. This implies that the electrode particle boundaries can still under tension or compression though no external load is applied. Without considering the interface effect, the radial stress is continuous at the interface since Eq. (23) reduces to the continuity condition and is automatically satisfied. Under the interface effect, a positive stress jump is observed at the interface. Please note the surface/interface effects are highlighted for this hollow biphase particle, due to its double surfaces and sharp interface.

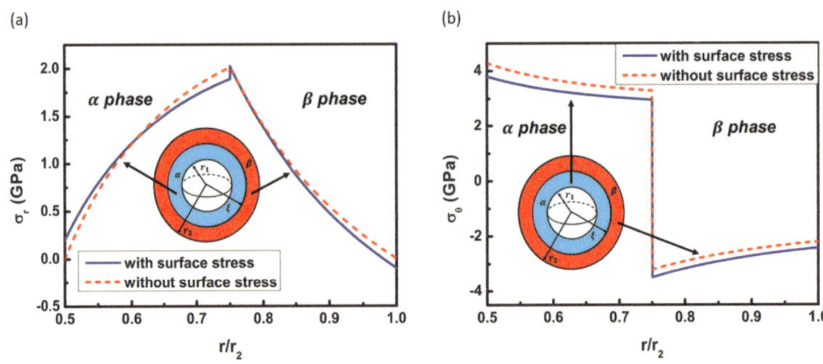

Figure 5. Stress field in the biphase electrode: (a) radial stress; (b) hoop stress (some data are reproduced from [47]).

Hoop stress is the driven force for the propagation of manufacture-induced imperfection. Therefore, hoop stress is more important in the stress analysis of electrode particles. One can observe that the hoop stress decreases from the inner surface to the outer one in the α phase. With a jump at the interface, the hoop stress changes from positive into negative value. It increases from phase interface to the outer surface. Compared to that of radial stress, the discontinuity of hoop stress at the interface is high. The reason for this high discontinuity is that the hoop stress is directly related to the interface discontinuous concentration. Therefore,

http://dx.doi.org/10.5772/62577

the effect of the phase transformation on hoop stress is worth more attention. By comparison, one can find that the surface/interface effect on hoop stress is much smaller.

The interactions between stress fields and phase transformation is systematically investigated in the above. One can conclude that the hydrostatic stress could accelerate the phase transformation and the time during phase transformation is overestimated by the conventional model without considering the effect of hydrostatic stress [33]. Moreover, the during the phase transformation, there is a sharp phase interface. The phase interface induced discontinuity in both radial and hoop stresses, and made the stress field in the electrode significantly higher. In summary, the phase transformation increases the stress field and therefore threatens the safety of electrode particle. But the hydrostatic stress could help the electrode to finish the phase transformation faster, and is important to avoid the failure of electrode particle.

5. Electrode geometry analysis

A key concept to avoid the stress-induced electrode failure is to find an optimal electrode, in which the stress is lower. In order to optimize the electrode, size and shape of the electrodes are two major topics to investigate. Different sized electrodes are studied during the last decades and considerable efforts are put into searching electrodes with optimal shapes, and researchers have tested multiple shapes [13, 34–36, 40–42]. However, those work mainly focus on the investigation of single phase solid electrodes. Hollow particle, with a unique doubled surface area, core-shell structure and large internal void, has great potentials in applications of lithium-ion batteries [42]. In Section 4, the concentration and stress profile of hollow electrode particles are analyzed. This section will study the size and shape effects of this hollow spherical particles.

5.1. Size effect

Size effect implies the dependence of the stress fields on the size of electrode particles, which has been investigated by researchers [11, 14, 18]. They conclude that larger particles lead to higher concentration gradient and induce higher stress. However, the model used in the above work is fairly simple. In the present model, the size effect can interact with the effects of hydrostatic stress and interface/surface stress, and have more complicated behaviors. To investigate this effect, a cross scale analysis is presented here, with particle size ranging from 10 nm to 20 μm and being discharged under 10 A/m^2 are simulated. The inner radii are half of the outer ones for these electrode particles.

5.1.1. Hydrostatic stress-induced size effect

Section 4 highlights the effect of hydrostatic stress on concentration and stress. In what follows, we will discuss the dependence of its effect on electrode size. **Figures 6a** illustrates the maximum hoop stresses of different sized electrode particles, when the concentration at the outer surface reaches c_{eq}^{α}. The stress fields with/without considering the effect of hydrostatic stress are compared. The stress fields predicted by two models are almost the same for the nanoelectrode particles, but are significantly different for the micro electrode particles. Without the effect of σ_h, large concentration gradient exists in large electrode particles, which induces the stress as high as 160 MPa. However, the stress field is not that high in reality. This is because σ_h can lower the concentration gradient and, therefore lower the stress fields. One can also explain the effect of σ_h as follows: the "effective" inward flux at outer surface $D\partial c/\partial r = i_n/F(1 + \theta c)$ decreases with c, which is equivalent to the case that the battery is discharged in a lower currency density. And it is well known that the stress of the electrode particle is lower when discharged under a lower currency density. Therefore, one can conclude that the effect of σ_h is necessary in stress predictions for the micro particles, but can be neglected for the nanoparticles.

5.1.2. Surface stress-induced size effect

In the community of solid mechanics, the surface stress-induced size effect is widely observed [45, 51, 52]. In 2008, Cheng et al. [11] introduces the surface stress effect to stress analysis of nano $LiMn_2O_4$ solid electrode particles in lithium-ion batteries. Here the size effect with surface stress is presented for the hollow spherical electrodes. We analyze the surface stress-induced size effect in the first stage of discharging process, since there is only α phase left and the effect of surface stress and phase transformation is decoupled. The maximum hoop stress $\sigma_{\theta max}$ of different-sized electrode particles are computed (c.f. **Figure 6b**). When the electrode particle is in micro size, the stresses predicted by the model with/without surface stress effect are almost the same. $\sigma_{\theta max}$ increases with particle size and reaches around 200 MPa when the particle size is 10 μm. However, a great difference is observed when the electrode particle size is nanoscale. Without considering the surface stress, $\sigma_{\theta max}$ is close to zero, which implies the whole particle is at the stress free state. But the stress predicted by the model with surface effect shows that the $\sigma_{\theta max}$ becomes negative, whose absolute values is large. This means the whole particle is under compression, which is important for the electrodes with flaw or small cracks, because the compression can stop the growth of these cracks. Therefore, the surface stress has to be considered in predicting the stress field of nanoelectrode particles.

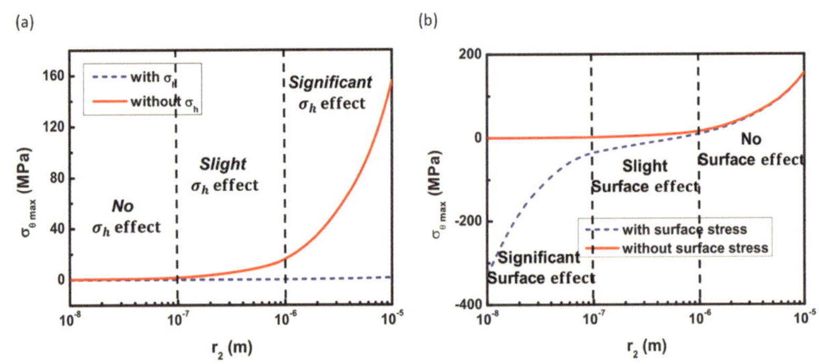

Figure 6. Variation of $\sigma_{\theta max}$ with size of electrode particles: (a) under effect of hydrostatic stress; (b) under effect of surface stress (some data are reproduced from [47]).

5.1.3. Interface stress-induced size effect

Although the surface effect of electrode particles has been reported for several years [11], the interface stress is not considered until the recent work of the authors [47]. In fact, for hollow electrode particles, the surface stress at the inner and outer surfaces are just special cases of interface stress Eq. (3). Since the size effect of surface stresses is reported above, the size effect due to interface stresses is expected.

Figures 7 illustrates the stress fields of different sized biphase electrode particles, that are discharged under 10 A/m^2 currency density. For the micro electrode particle, the interface stress is not significant and the discontinuity of σ_r at phase interface is small. However, when the size of electrode reaches nanoscale, the interface stress is much higher, which leads to a higher stress discontinuity at the phase interface. Recall that before the phase transformation, the nanoelectrode particles is under compression due to the surface stress. But it is under tension due to the high stress discontinuity. Because the tensional stress may drive the existing crack to propagate, the biphase stage is the most dangerous. Furthermore, stresses increase by more than 200 times in this stage, and threaten the safety of electrodes.

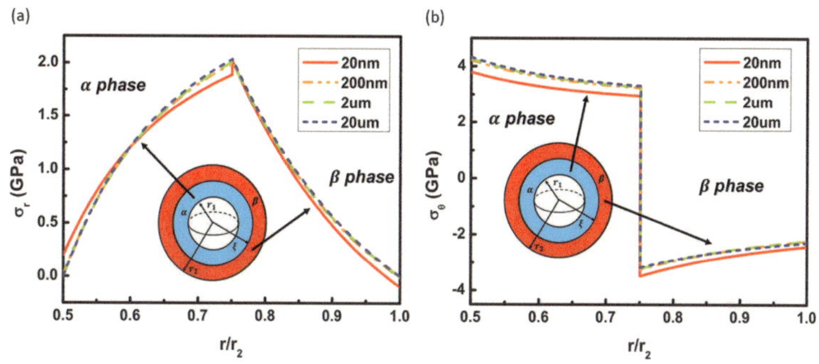

Figure 7. Variation of stress fields with sizes of electrodes: (a) σ_r; (b) σ_θ (some data are reproduced from [47]).

In summary, hydrostatic stress, surface stress, and interface stress can all induce size effect. The effect of hydrostatic stress increases with particle size while the effects of surface and interface stresses decrease with particle size. This is because the hydrostatic stress is inserted into bulk equations while the surface/interface stress is considered in the surface/interface equations. The ratio of surface/interface over volume decreases with the particle size, so do the effects of surface/interface stresses.

5.2. Shape effect

Because of its better cyclability, higher rate capability, and less capacity degradation compared to the solid sphere electrodes [46, 48, 49], hollow electrode particle attracts wide research attention. The structural feature of hollow particles, i.e., shape effect, can be characterize by a parameter $\zeta = r_1/r_2$. The hollow sphere could reduce to solid ones when $\zeta = 0$ ($r_1 = 0$), and become a thin shell when $\zeta \approx 1$ ($r_1 \approx r_2$). The shape effect on the mechanical and electrochemical properties of hollow electrode particles are systematically investigated in the following.

5.2.1. Shape effect on stress

The maximum and minimum hoop stresses of hollow particles with different ζ but same volume are compared (**Figure 8a**). To highlight this shape effect, no interface/surface and hydrostatic stresses are considered here. One can observe that the absolute values of maximum and minimum hoop stresses both decrease with ζ. The stress of the particles with $\zeta = 0.05$ is even 20 times more than that with $\zeta = 0.95$. The advantages of hollow electrode particles will be strengthened in thin shell structure, and zero stress state is expected when $\zeta \to 1$. Stress profiles are calculated for the hollow particles with same outer radius 13.82 μm, but with different inner radii. The maximum and minimum hoop stresses ($\sigma_{\theta max}$ and $\sigma_{\theta min}$) are illustrated in **Figure 8b**. Observations show that the absolute values of $\sigma_{\theta max}$ and $\sigma_{\theta min}$ decrease with ζ. In other words, the stresses decrease with the shell thickness of hollow particles. When $\zeta \to 1$, the thickness is close to zero and the hollow particle approaches a single thin layer shell in which no concentration gradient or stress exist.

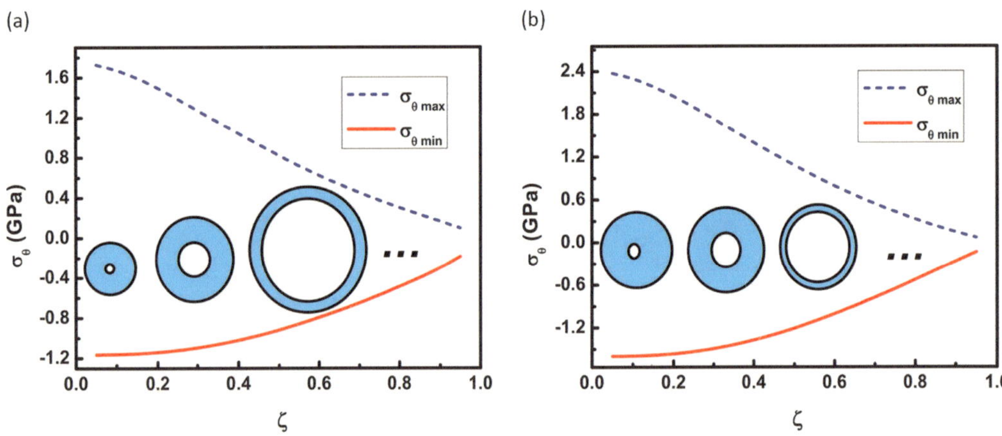

Figure 8. Maximum and minimum hoop stresses: (a) different sizes with same volume; (b) different volumes with same sizes (some data are reproduced from [47]).

In conclusion, the hollow structure electrodes can significantly reduce the stress field of electrode. The hoop stress decreases with the thickness of the shell. Furthermore, because the tensional hoop stress is the driven force of crack growth in the electrode, hollow structure can help to lower the failure possibility of lithium ion batteries.

5.2.2. Shape effect in efficiency

At the end of galvanostatic process, there is some residual capacity. The residual lithium will be in waste without the potentiostatic operation. To characterize the effective capacity of the battery, a variable η is defined as the ratio of effective capacity to total capacity,

$$\eta = \frac{\mathrm{Cap_t} - \mathrm{Cap_{re}}}{\mathrm{Cap_t}} \tag{37}$$

where $\mathrm{Cap_t}$ and $\mathrm{Cap_{re}}$ are total and residual capacities. Please note that $\eta \le 1$ since $\mathrm{Cap_{re}} \ge 0$. By performing the potentiostatic charging, the battery gets fully charged/discharged, i.e., $\mathrm{Cap_{re}}$ can reduce to 0 and η will reach 1.

For this special electrode, the total capacity is

$$\mathrm{Cap_t} = \frac{4\pi}{3} Q\rho\Omega c_{max}(r_2^3 - r_1^3) \tag{38}$$

where Q is the theoretical capacity of the electrode material. The residual capacity of the electrode particle is

$$\mathrm{Cap_{re}} = 4\pi Q\rho\Omega \int_{r_1}^{r_2} cr^2 dr \tag{39}$$

Substitution of Eq. (39) into Eq. (38) yields

$$\eta = 1 - \frac{3\int_{r_1}^{r_2} cr^2 dr}{c_{\max}\left(r_2^3 - r_1^3\right)} \tag{40}$$

$Li_xMn_2O_4$ is taken as an example to analyze the efficiency of the electrodes, whose material properties are provided in [14]. Variation of η with ζ is illustrated (Figure 9). One can observe that η increases with ζ, and approaches 1 when ζ is close to 1. This implies that the effective capacity can be high enough without potentiostatic charging as long as the shell is thin enough.

The particle sizes and charging densities have significant influence on the stress profile of electrode particles. In what follows, their effects on the efficiency are investigated. **Figure 9a** illustrates the dependence of efficiency on charging currency density when the electrode particles have the same size r_2 = 20 μm. Observations show that η becomes higher when the electrode is charged under lower currency density. The efficiency η is above 96% regardless of ζ, when $i_n = 5$ A/m^2. The efficiencies of different-sized electrode particles are compared in **Figure 9b**. The particles are all charged under the same currency density $i_n = 20$ A/m^2. One can observe that smaller electrode particles have higher efficiency. When the electrode particle is small (i.e. r_2 = 5 μm), the efficiency is above 96%.

In summary, the efficiency of electrode particle depends on particle size, shell thickness, and charging currency density. High effective capacity can be obtained by using thin shell hollow particle, small electrode particle, and low currency density.

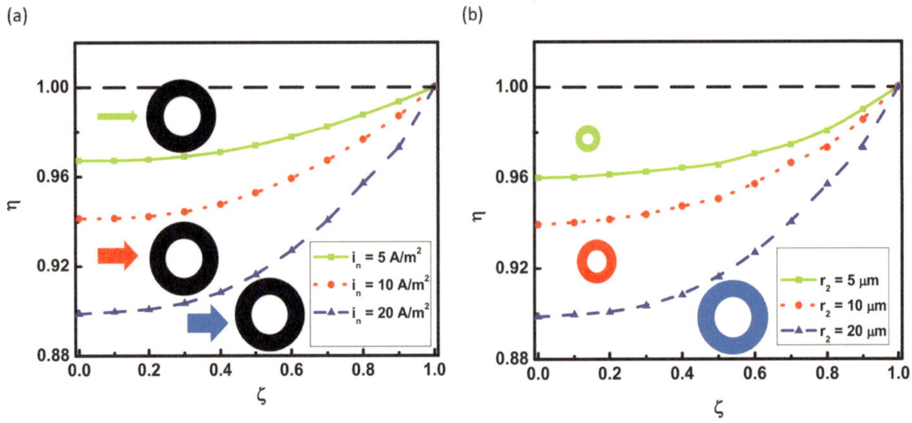

Figure 9. Variation of η with ζ: (a) different currency densities; (b) different sizes (some data are reproduced from [47]).

The size and shape effect of hollow spherical particles are systematically investigated above. The size effect is observed for the hydrostatic stress, surface stress, and interface stress, which implies that hydrostatic stress is important in predicting the stress of micro particles while surface and interface stresses are essential in computing the stress fields of nanoparticles. The advantages of hollow electrode particles are highlight in significantly reducing the stress and increasing the efficiency. In summary, hollow structure particle can significantly improve the mechanical and electrochemical properties and is a good candidate for the optimal electrode particle.

6. Conclusions

This chapter reviews the models on stress analysis of electrode particles in lithium-ion batteries, and then provides an electrochemo-mechanical framework to model the concentration and stress of an electrode with arbitrary geometry. The hydrostatic stress and surface/interface stress are considered. The interactions between phase transformation and stress profile are investigated. The equations are then reformulated for the specific hollow spherical electrode particles. Conclusions are summarized as follows:

1. The interaction between stress fields and phase transformation is fully addressed. Under the effect of hydrostatic stress, the phase transformation process is much faster, which implies the time of phase transformation is overestimated in previous publications. Due to the existence of phase interface, the stress becomes discontinuous at the interface. The stresses filed of the whole electrode is much larger in the biphase stage.

2. The size and shape effects for hollow spherical particle are investigated. Through a cross-scale analysis, we conclude that hydrostatic stress, surface stress and interface stress can all induce size effect. Hydrostatic stress is important in predicting the stress of micro particles while surface and interface stresses are essential in computing the stress fields of nanoparticles.

3. The shape effect of the hollow electrode particle is analyzed, which highlights its structural advantages. One can find that the stress field highly depends on ζ (=r_1/r_2). The maximum hoop stress decreases with ζ. When ζ is close to 1, the stress is approaching zero. A variable η, to characterize the efficiency (ratio of effective capacity over total capacity) of battery is defined. η also decreases with ζ. When the thickness of hollow electrode particle is small enough, η is close to 1. Therefore, the hollow particle is a good candidate for the optimal electrode particle.

This chapter presents an electrochemo-mechanical framework to accurate predict the stress profile in the electrode particles with different sizes and geometries. The proposed framework considers different effects simultaneously, and quantitatively compares their contributions. This chapter also proposes an optimal electrode particle, i.e., the hollow spherical particle, and systematically analyzes its size and shape effects. Therefore, the present chapter is helpful for the material and structure design of electrode.

Acknowledgements

This work is sponsored by the Alexander von Humboldt (AvH) foundation through project 'Mechanics theory of materials with complex surfaces and its applications'; Huiling Duan appreciates the support of following agencies: Major State Basic Research Development Programme of China, National Natural Science Foundation of China.

Author details

Yingjie Liu[1] and Huiling Duan[2*]

*Address all correspondence to: hlduan@pku.edu.cn

1 Computational Mechanics Lab, Department of Civil and Environmental Engineering, Duke University, Durham, NC, USA

2 State Key Lab for Turbulence and Complex System, Department of Mechanics and Engineering Science, Peking University, Beijing, PR China

References

[1] Baker DR, Verbrugge MW. The role of charge separation in the response of electrochemical systems. Proc. R. Soc. Lond. A. 2002;454,1805–1829. (DOI:10.1098/rspa.1998.0233)

[2] Yi Y-B, Sastry AM. Analytical approximation of the percolation threshold for overlapping ellipsoids of revolution. Proc. R. Soc. Lond. A. 2004;460,2353–2380. (DOI:10.1098/rspa.2004.1279)

[3] Wu Y, Pasero D, Mc Cabe EE, Matsushima Y, West AR. Partial cation-order and early-stage, phase separation in phase W, $Li_xCo_{1-x}O$: $0.07 \le x \le 0.24 - 0.3$. Proc. R. Soc. A. 2009;465,1829–1841. (DOI:10.1098/rspa.2008.0489)

[4] Mc Kenna KP, Shluger AL. Electron and hole trapping in polycrystalline metal oxide materials. Proc. R. Soc. A. 2011;467,2043–2053. (DOI:10.1098/rspa.2010.0518)

[5] Maksimchuk NV, Zalomaeva OV, Skobelev IY, Kovalenko KA, Fedin VP, Kholdeeva OA. Metal-organic frameworks of the MIL-101 family as heterogeneous single-site catalysts. Proc. R. Soc. A. 2012;468,2017–2034. (DOI:10.1098/rspa.2012.0072)

[6] Chen CQ, et al. Perspectives in mechanics of heterogeneous solids. Acta Mech. Solida Sin. 2011;24,1–26. (DOI:10.1016/S0894-9166(11)60007-4)

[7] Christensen J, Newman J. A mathematical model of stress generation and fracture in lithium manganese oxide. J. Electrochem. Soc. 2006;153,A1019–A1030. (DOI: 10.1149/1.2185287)

[8] Jang YI, Huang BY, Wang HF, Sadoway DR, Ceder G, Chiang YM, Liu H, Tamura H. $LiAlyCo_{1-y}O_2$ (R_{3m}) intercalation cathode for rechargeable lithium batteries. J. Electrochem. Soc. 1999;146,862–868. (DOI:10.1149/1.1391693)

[9] Wang HF, Jang Y-II, Huang B, Sadoway DR, Chiang Y-M. Electron microscopic characterization of electrochemically cycled $LiCoO_2$ and $Li(Al,Co)O_2$ battery cathodes. J. Power Sources. 1999;81–82,594–598. (DOI:10.1016/S0378-7753(99)00108-1)

[10] Gabrisch H, Wilcox J, Doeff MM. TEM study of fracturing in spherical and plate-like $LiFePO_4$ particles. Electrochem. Solid State Lett. 2008;11,A25–A29. (DOI: 10.1149/1.2826746)

[11] Cheng Y-T, Verbrugge MW. The influence of surface mechanics on diffusion induced stresses within spherical nanoparticles. J. Appl. Phys. 2008;104,083521. (DOI: 10.1063/1.3000442)

[12] Cheng Y-T, Verbrugge MW. Evolution of stress within a spherical insertion electrode particle under potentiostatic and galvanostatic operation. J. Power Sources. 2009;190,453–460. (DOI:10.1016/j.jpowsour.2009.01.021)

[13] Lu B, Song YC, Guo ZS, Zhang JQ. Analysis of delamination in thin film electrodes under galvanostatic and potentiostatic operations with Li-ion diffusion from edge. Acta. Mech. Sinica. 2013;29,348–356. (DOI: 10.1007/s10409-013-0038-x)

[14] Woodford WH, Chiang Y-M, Carter WC. 'Electrochemical shock' of intercalation electrodes: a fracture mechanics analysis. J. Electrochem. Soc. 2010;157,A1052–A1059. (DOI:10.1149/1.3464773)

[15] Hu YH, Zhao XH, Suo ZG. Averting cracks caused by insertion reaction in lithium-ion batteries. J. Mater. Res. 2010;25,1007–1010. (DOI:10.1557/JMR.2010.0142)

[16] Zhao KJ, Wang WL, Gregoire J, Pharr M, Suo ZG, Vlassak JJ, Kaxiras E. Lithium-assisted plastic deformation of silicon electrodes in lithium-ion batteries: a first-principles theoretical study. Nano Lett. 2011;11,2962–2967. (DOI:10.1021/nl201501s)

[17] Zhang XC, Sastry AM, Shyy W. Intercalation-induced stress and heat generation within single lithium-ion battery cathode particles. J. Electrochem. Soc. 2008;155,A542–A552. (DOI:10.1149/1.2926617)

[18] Zhang XC, Shyy W, Sastry AM. Numerical simulation of intercalation-induced stress in li-ion battery electrode particles. J. Electrochem. Soc. 2007;154,A910–A916. (DOI: 10.1149/ 1.2759840)

[19] Park J, Lu W, Sastry AM. Numerical simulation of stress evolution in lithium manganese dioxide particles due to coupled phase transition and intercalation. J. Electrochem. Soc. 2011;158,A201–A206. (DOI:10.1149/1.3526597)

[20] Park J, Seo JH, Plett G, Lu W, Sastry AM. Numerical simulation of the effect of the dissolution of $LiMn_2O_4$ particles on li-ion battery performance. Electrochem. Solid-State Lett. 2011;14,A14–A18. (DOI:10.1149/1.3516619)

[21] Cheng Y-T, Verbrugge MW. Application of hasselman's crack propagation model to insertion electrodes. Electrochem. Solid-State Lett. 2010;13,A128–A131. (DOI: 10.1149/1.3455179)

[22] Zhao KJ, Pharr M, Vlassak JJ, Suo ZG. Fracture of electrodes in lithium-ion batteries caused by fast charging. J. Appl. Phys. 2010;108,073517. (DOI:10.1063/1.3492617)

[23] Grantab R, Shenoy VB. Location- and orientation-dependent progressive crack propagation in cylindrical graphite electrode particles. J. Electrochem. Soc. 2011;158,A948–A954. (DOI:10.1149/1.3601878)

[24] Haftbaradaran H, Gao HJ, Curtin WA. A surface locking instability for atomic intercalation into a solid electrode. Appl. Phys. Lett. 2010;96,091909. (DOI: 10.1063/1.3330940)

[25] Haftbaradaran H, Song J, Curtin WA, Gao HJ. Continuum and atomistic models of strongly coupled diffusion, stress, and solute concentration. J. Power Sources. 2011;196,361–370. (DOI:10.1016/j.jpowsour.2010.06.080)

[26] Hu YH, Zhao XH, Suo ZG. Averting cracks caused by insertion reaction in lithium-ion batteries. J. Mater. Res. 2010;25,1007–1010. (DOI:10.1557/JMR.2010.0142)

[27] Zhao KJ, Wang WL, Gregoire J, Pharr M, Suo ZG, Vlassak JJ, Kaxiras E. Lithium-assisted plastic deformation of silicon electrodes in lithium-ion batteries: a first-principles theoretical study. Nano Lett. 2011;11,2962–2967. (DOI:10.1021/nl201501s)

[28] Nytén A, Abouimrane A, Armand M, Gustafsson T, Thomas JO. Electrochemical performance of Li_2FeSO_4 as a new li-battery cathode material. Electrochem. Commun. 2005;7,156–160. (DOI:10.1016/j.elecom.2004.11.008)

[29] Ramesh TN, Lee KT, Ellis BL, Nazar LF. Tavorite lithium iron fluorophosphate cathode materials: phase transition and electrochemistry of $LiFePO_4F$-Li_2FePO_4F. Electrochem. Solid-State Lett. 2010;13,A43–A47. (DOI:10.1149/1.3298353)

[30] Shin HC, Pyun SI. An investigation of the electrochemical intercalation of lithium into a $Li1{-}\delta\ CoO_2$ electrode based upon numerical analysis of potentiostatic current transients. Electrochim. Acta. 1999;44,2235–2244. (DOI:10.1016/S0013-4686(98)00340-5)

[31] Shin HC, Pyun SI. The kinetics of lithium transport through $Li1{-}\delta\ CoO_2$ by theoretical analysis of current transient. Electrochim. Acta. 1999;45,489–501. (DOI:10.1016/S0013-4686(99)00270-4)

[32] Zhang Q, White RE. Moving boundary model for the discharge of a $LiCoO_2$ electrode. J. Electrochem. Soc. 2007;154,A587–A596. (DOI:10.1149/1.2728733)

[33] Renganathan S, Sikha G, Santhanagopalan S, White RE. Theoretical analysis of stresses in a lithium ion cell. J. Electrochem. Soc. 2010;157,A155–A163. (DOI:10.1149/1.3261809)

[34] Li JC-M. Physical chemistry of some microstructural phenomena. MTA. 1978;9,1353–1380. (DOI:10.1007/BF02661808)

[35] Hao F, Fang DN. Diffusion-induced stresses of spherical core-shell electrodes in lithium- ion batteries: the effects of the shell and surface/interface stress. J. Electrochem. Soc. 2013;160,A595–A600. (DOI:10.1149/2.054304jes)

[36] Hao F, Fang DN. Tailoring diffusion-induced stresses of core-shell nanotube electrodes in lithium-ion batteries. J. Appl. Phys. 2013;113,013507. (DOI:10.1063/1.4772963)

[37] Deshpande R, Cheng Y-T, Verbrugge MW. Modeling diffusion-induced stress in nanowire electrode structures. J. Power Sources. 2010;195,5081–5088. (DOI:10.1016/j.jpowsour. 2010.02.021)

[38] Xiao X, Liu P, Verbrugge MW, Haftbaradaran H, Gao HJ. Improved cycling stability of silicon thin film electrodes through patterning for high energy density lithium batteries. J. Power Sources. 2011;196,1409–1416. (DOI:10.1016/j.jpowsour.2010.08.058)

[39] Lu B, Song YC, Guo ZS, Zhang JQ. Modeling of progressive delamination in a thin film driven by diffusion-induced stresses. Int. J. Solids Struct. 2013;50,2495–2507. (DOI: 10.1016/j.ijsolstr.2013.04.003)

[40] Zhang JQ, Lu B, Song YC, Ji X. Diffusion induced stress in layered Li-ion battery electrode plates. J. Power Sources. 2012;209,220–227. (DOI:10.1016/j.jpowsour. 2012.02.104)

[41] Bhandakkar TK, Gao HJ. Cohesive modeling of crack nucleation under diffusion induced stresses in a thin strip: implications on the critical size for flaw tolerant battery electrodes. Int. J. Solids Struct. 2010;47,1424–1434. (DOI:10.1016/j.ijsolstr.2010.02.001)

[42] Bhandakkar TK, Gao HJ. Cohesive modeling of crack nucleation in a cylindrical electrode under axisymmetric diffusion induced stresses. Int. J. Solids Struct. 2011;48,2304–2309. (DOI:10.1016/j.ijsolstr.2011.04.005)

[43] Keng PY, et al. Colloidal polymerization of polymer-coated ferromagnetic nanoparticles into cobalt oxide nanowires. ACS Nano. 2009;3,3143–3157. (DOI:10.1021/nn900483w)

[44] Duan HL, Wang J, Huang ZP, Zhong Y. Stress fields of a spheroidal inhomogeneity with an interphase in an infinite medium under remote loadings. Proc. R. Soc. A. 2005;461,1055–1080. (DOI:10.1098/rspa.2004.1396)

[45] Duan HL, Weissmueller J, Wang Y. Instabilities of core-shell heterostructured cylinders due to diffusions and epitaxy: spheroidization and blossom of nanowires. J. Mech. Phys. Solids. 2008;56,1831–1851. (DOI:10.1016/j.jmps.2007.11.009)

[46] Yao Y, Mc Dowell MT, Ryu I, Wu H, Liu N, Hu LB, Nix WD, Cui Yi. Interconnected silicon hollow nanospheres for lithium-ion battery anodes with long cycle life. Nano Lett. 2011;11,2949–2954. (DOI:10.1021/nl201470j)

[47] Liu YJ, Lv PY, Ma J, Bai RB, Duan HL. Stress fields in hollow coreshell spherical electrodes of lithium ion batteries. Proc. R. Soc. Lond. A. 2014;470,20140299. (DOI: 10.1098/rspa.2014.0299)

[48] Yoon S, Manthiram A. Hollow Core-Shell mesoporous TiO_2 spheres for lithium ion storage. J. Phys. Chem. C. 2011;115, 9410–9416. (DOI:10.1021/jp1123184)

[49] Koo B, Xiong H, Slater MD, Prakapenka VB, Balasubramanian M, Podsiadlo P, Johnson CS, Rajh T, Shevchenko EV. Hollow iron oxide nanoparticles for application in lithium ion batteries. Nano Lett. 2012;12, 2429–2435. (DOI:10.1021/nl3004286)

[50] Lu, B., Song, Y. C., Guo, Z. S., Zhang, J. Q. Modeling of progressive delamination in a thin film driven by diffusion-induced stresses. Int. J. Solids Struct. 2013;50, 2495–2507. (DOI:10.1016/j.ijsolstr.2013.04.003)

[51] Duan HL, Wang J, Huang ZP, Karihaloo BL. Eshelby formalism for nano- inhomogeneities. Proc. R. Soc. A. 2005;461, 3335–3353. (DOI:10.1098/rspa.2005.1520)

[52] Weissmüller J, Duan HL, Farkas D. Deformation of solids with nanoscale pores by the action of capillary forces. Acta Mater. 2010;58, 1–13. (DOI:10.1016/j.actamat.2009.08.008)

[53] Yvonnet J, Le-Quang H, He QC. An XFEM/level set approach to modeling surface/interface effects and to computing the size-dependent effective properties of nanocomposites. Comput. Mech. 2008;42, 119–131. (DOI:10.1007/s00466-008-0241-y)

[54] Yvonnet J, Le-Quang H, Toulemonde C, He QC. Thermo-mechanical modelling of materials containing micro/nano inclusions with imperfect interfaces. Int. J. Mater. Form. 2008;1, 1139–1142. (DOI:10.1007/sl2289-008-0181-2)

[55] Wang WL, Lee S, Chen JR. Effect of chemical stress on diffusion in a hollow cylinder. J. Appl. Phys. 2002;91, 9584–9590. (DOI:10.1063/1.1477624)

[56] Fischer FD, Svoboda J. Stresses in hollow nanoparticles. Int. J. Solids Struct. 2010; 47, 2799–2805. (DOI:10.1016/j.ijsolstr.2010.06.008)

Section 2

Other Alkali-ion Batteries

Chapter 8

High-Voltage Cathodes for Na-Ion Batteries: Sodium–Vanadium Fluorophosphates

Paula Serras, Verónica Palomares and Teófilo Rojo

Additional information is available at the end of the chapter

http://dx.doi.org/10.5772/62317

Abstract

This chapter analyses the main advances made in the field of sodium–vanadium fluorophosphates as cathodes for Na-ion batteries and tries to clarify some discrepancies and common errors published about these compounds. The sodium–vanadium fluorophosphate family can be divided in two main members: $Na_3V_2(PO_4)_2F_3$ (V^{+3} extreme phase) and $Na_3V_2O_2(PO_4)_2F$ (V^{4+} extreme phase). $Na_3V_2O_{2x}(PO_4)_2F_{3-2x}$, where $0<x<1$ would correspond to intermediate $V^{3+/4+}$ mixed valence phases. Among them, the V^{3+} extreme has demonstrated to be difficult to isolate, whereas the V^{4+} and mixed valence phases can be more easily prepared by different synthesis methods and from different vanadium sources. In terms of electrochemical performance, mixed valent compound provides good performance, with high specific capacity at moderate/high cycling rates, and long cycle life. The future perspectives for this family of compounds are discussed in terms of raw materials availability, price, and performance relative to other cathode systems for Na-ion batteries.

Keywords: Na-ion batteries, cathodes, fluorophosphates, vanadium, High voltage

1. Introduction

Energy production is a key issue for the support of our society and way-of-life. Driven by the need to reduce the emissions of CO_2 and increase energy security, policy makers have implemented different measures in order to shift to low-carbon energy resources. One of these policies is to promote the use of renewable energy sources at the expense of the carbon-based fuels [1,2]. However, the increasing use of these clean energy sources entails other problems, such as modulating time-variable energy production from renewable resources due to their dependence on the weather to integrate them into the grid. Thus, in order to solve the problem concerning the intermittency of these resources, additional energy storage devices are needed. This way, electrochemical energy storage is a strategic research area that will help to increase the weight of renewable energy sources in the energy mix.

In this field, batteries are one of the most promising systems to act as buffer to regulate the variable energy income that can be obtained from the wind, sun and from the ocean waves. For this purpose, one of the main technologies under development is Na-ion batteries. In the last 20 years, Li-ion batteries have attracted all the attention in energy storage, but recently, Na-based compounds have made a comeback because of controversial debates regarding the size of reserves and higher cost to obtain Li [3]. The use of Na instead of Li in rocking chair batteries could mitigate the feasible shortage of lithium in an economic way for many reasons. The first one, the unlimited sodium sources, being the sodium content in the earth's crust and water of 28,400 $mg{\cdot}kg^{-1}$ and 11,000 $mg{\cdot}L^{-1}$ compared to 20 $mg{\cdot}kg^{-1}$ and 0.18 $mg{\cdot}L^{-1}$ for lithium [4]. Moreover, sodium presents lower price than lithium and it is easy to recover [5].

Although sodium (Na)-ion batteries are considered a promising alternative to lithium-ion systems, it is usually said that they possess limited electrochemical activity compared to Li due to two intrinsic reasons. First, Na has a lower ionization potential than Li (5.139eV vs. 5.392 eV) [6], leading to lower operating voltages and thus lower energy densities. Second, Na^+ ions are heavier and larger than Li^+ ions, leading to slow diffusion within a solid electrode during cycling and larger volume expansion of the electrode [7]. These two factors would induce lower gravimetric energy densities, so renewed research on these systems has been focused on their utility as buffering systems in the electrical grid, where battery volume is not so crucial due to the stationary nature of the practical use of these batteries.

However, and in spite of the great efforts devoted to the development of higher performance Li-ion batteries, if practical energy densities achieved up till now for both Li- and Na-based systems are compared, it can be seen that Na-ion batteries are catching up with the practical energy densities of Li-ion in a quite short time (**Figure 1**).

This great advance in gravimetric energy densities for Na-ion batteries is due to the extensive research that has been done in the last years to identify and optimise the best materials for these systems, so they can be soon commercially viable. As it has been said, the incorporation of this kind of systems to the electrical grid would lead to a significant increase of the proportion of renewable energies on the energy mix, with the positive environmental consequences that this fact would result in. Thus, it is a priority to continue searching for and optimising the best-performing materials to be used in the different components of a battery: cathode, electrolyte and anode.

In the case of the cathode, a good performing material should provide high energy density to the battery, that is, it should be able to charge and discharge the highest possible specific capacities with also high operating voltage and for many cycles (long cycle life). In this sense, sodium–vanadium fluorophosphates have demonstrated to fulfil all of these characteristics.

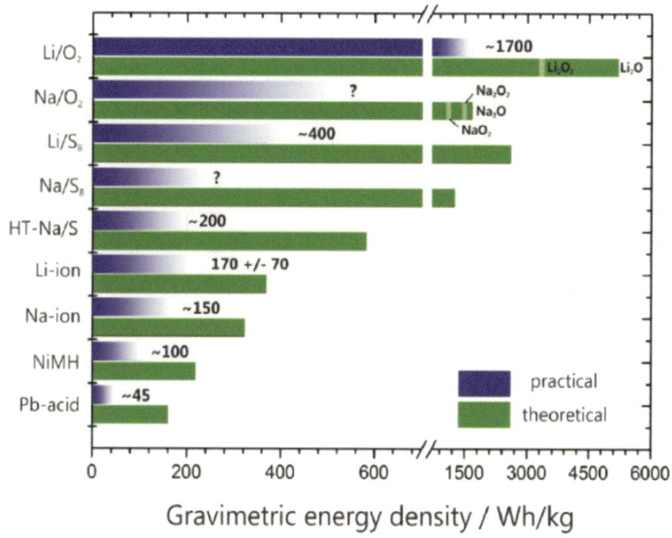

Figure 1. Theoretical and (estimated) practical energy densities of different rechargeable batteries: Pb–acid (lead acid), NiMH (nickel metal hydride), Na-ion (estimate derived from data for Li-ion assuming a slightly lower cell voltage), Li-ion (average over different types), HT-Na/S_8 (high temperature sodium–sulphur battery), Li/S_8 and Na/S_8 (lithium–sulphur and sodium–sulphur battery assuming Li_2S and Na_2S as discharge products), Li/O_2 and Na/O_2 (lithium–oxygen battery, where theoretical values include the weight of oxygen and depend on the stoichiometry of the assumed discharge product, i.e., oxide, peroxide or superoxide). Reproduced from reference [8].

2. High-voltage cathode materials for Na-ion batteries

In the last decade, three dimensional frameworks built on transition metals (M) and polyanions $(XO_4)^{n-}$ have become a hot topic in the research field of electroactive materials for lithium and Na-ion batteries. The smaller theoretical gravimetric capacity achieved due to the presence of polyanion groups is compensated by the positive features presented by these compounds, such as very stable frameworks and high inductive effect [9]. The "inductive effect" is a concept proposed by Goodenough *et al.* consisting on the tune of the redox potential of the 3d metal centre (M) depending on the electronegativity of the X-O bonding [10].

On one hand, sulphate-based cathodes offer a good combination of sustainable syntheses and high energy density owing to their high-voltage operation due to electronegative SO_4^{2-} units [11]. $Na_2Fe_2(SO_4)_3$ has demonstrated to be the most promising compound being operable at 3.8 V *versus* sodium and offering a reversible capacity of over 100 mAhg^{-1} with a 50% capacity retention at high rates (20C) [12]. However, the main issues concerning sulphur-containing cathodes are the low utilisation of active material, volume expansion during extraction/insertion of Na ions and the dissolution of polysulphide into the electrolyte [13].

On the other hand, framework materials based on the phosphate polyanion have also been identified as promising electro-active materials for sodium metal and Na-ion battery applications. It is the strong inductive effect of the PO_4^{3-} polyanion that moderates the energetic of the transition metal redox couple to generate relatively high operating potentials for these compounds [14]. $NaMnPO_4$ is one of the most studied compounds in this field due to the high operating voltage provided by its lithium analogue: $LiMnPO_4$ which works at 4.1 V *versus* lithium [15]. However, contrary to $LiMnPO_4$, $NaMnPO_4$ possesses two structure modifications: marycite-type and olivine-type structures being this latter one the electroactive one. Boyadzhieva *et al.* have recently reported, for the first time, the reversible lithium and Na ion intercalation into the olivine $NaMnPO_4$ phase when testing it *versus* lithium. However, the electrochemical performance of this material is not yet competitive compared to the well-known $LiMnPO_4$ analogue so that further optimisation is needed [16].

$Na_4Co_3(PO_4)_2P_2O_7$ pyrophosphate is another promising candidate due to its high working potential region (between 4.1 and 4.7 V *versus* sodium). In the work presented by Nose *et al.* [17], this compound offers 95 $mAhg^{-1}$ reversible capacity with negligible fading after 100 cycles and 84% of capacity retention at rates of 25C. Moreover, the battery presented may be capable of running at even higher potentials (~5.1 V) in order to utilise the $Co^{3+/4+}$ redox couple leading to the extraction of 4 Na ions from the $Na_4Co_3(PO_4)_2P_2O_7$ phase. Something similar is proposed by Xu *et al.* in the case of $Na_3V_2O_2(PO_4)_2F$ sodium–vanadium fluorophosphate. Due to the difficulty to access the third sodium in the formula unit some theoretical calculations suggested the substitution of oxygen with chlorine to form $Na_3V_2Cl_2(PO_4)_2F$ which would increase the energy density from 520 to 758 $Whkg^{-1}$[18]. However, a suitable electrolyte would be needed in both cases in order to work in such high-voltage conditions.

Among the framework materials fluorophosphates possess even higher operating voltages than phosphates, because the inductive effect of fluorine is added to the effect of phosphate. This latter feature makes them a key to solve the energy density issue of sodium-based batteries. Fe, Mn, and V have been the most investigated transition metals but the three of them present different structures: Whereas the sodium–iron fluorophosphate possesses a two-dimensional layered structure [19], the sodium–manganese fluorophosphate presents a three-dimensional tunnel structure [20]. However, manganese compound has demonstrated to be poorly electrochemically active so that more studies have been performed for the iron-based compound. For the moment, one of the best results achieved has been for carbon coated porous hollow spheres of Na_2FePO_4F phase. This nanostructured material contained about 6–8 *wt.*% of carbon and presented a specific surface area of 8.7 m^2g^{-1} which led to a specific capacity of 90 $mAhg^{-1}$ at C/10 and a good cycling stability for 100 cycles [21].

Regarding sodium–vanadium fluorophosphates, three phases have been described in the literature: $NaVPO_4F$, $Na_3V_2O_2(PO_4)_2F$ and $Na_3V_2(PO_4)_2F_3$ with theoretical specific capacity values of 143, 130 and 128 $mAh{\cdot}g^{-1}$, respectively. Barker *et al.* first proposed $NaVPO_4F$ as a tetragonal structure with *I4/mmm* space group [22]. On the other hand, several groups reported also the existence of a $NaVPO_4F$ polymorph which they indexed with *C2/c* space group [23–26]. The $Na_3V_2O_2(PO_4)_2F$ phase has been described as a tetragonal symmetry compound with *I4/mmm* space group by Sauvage *et al.* [27] and Massa *et al.* [28] but, more recently, Tsirlin *et al.* [29] proposed the $P4_2/mnm$ space group for the room temperature phase among the diverse polymorphs of this material at different temperatures. The third of the sodium–vanadium fluorophosphates mentioned in the literature, $Na_3V_2(PO_4)_2F_3$, is also described as a tetragonal

http://dx.doi.org/10.5772/62317

symmetry with *$P4_2/mnm$* space group [30–32] but more recently some authors have proposed a different space group for this compound: *Amam* [33,34].

A deep study of the bibliographic data related to these three sodium fluorophosphates leads to doubt about the real existence of these three different compounds. It is worth noting that the existence of the $NaVPO_4F$ phase has already been questioned by authors such as Sauvage *et al.* [27]. Furthermore, there are no structural data about the suggested monoclinic polymorph of $NaVPO_4F$ [23–26]. Besides, the diffractogram shown in their works as well as the electrochemical curves obtained match with the one of $Na_3V_2(PO_4)_3$ NASICON compound. In this case, the formation of a NASICON type phase instead of $NaVPO_4F$ could be possible because the samples were prepared by the ceramic method and sublimation of VF_3 could occur, as it has been described for other lithium–vanadium fluorophosphates [35]. Thus, structural data of only two compounds are found in the literature: $Na_3V_2O_2(PO_4)_2F$ [29] and $Na_3V_2(PO_4)_2F_3$ [30]. The diffractograms of both phases are quite similar that it is possible to establish a relationship between them.

Electrochemical studies on the three mentioned tetragonal structure compounds have been performed under different conditions *versus* lithium or sodium [22–24,27,36–39] and in all cases the charge–discharge curves display two voltage *plateaux* of similar length at the same voltages with little differences depending on the anode material used (*ca.* 3.6 and 4.1 V *versus* Na/Na^+).

3. Sodium–vanadium fluorophosphates as cathode materials for Na-ion batteries

The study of sodium–vanadium fluorophosphates is especially relevant because of the high operating voltages offered by these compounds that could lead the way to high energy Na-ion batteries.

The similarity of the X-ray diffraction patterns as well as the almost identical electrochemical data for $Na_3V_2O_2(PO_4)_2F$ and $Na_3V_2(PO_4)_2F_3$ suggested that both materials could belong to the same family of compounds, where the fluorine content is modulated by the presence of V^{3+} and VO^{2+} (V^{4+}) leading to the following general formula $Na_3V_2O_{2x}(PO_4)_2F_{3-2x}$ which was proposed by our group for the first time [40]. The extreme members ($x = 0$ and $x = 1$) would correspond to the mentioned phases, $Na_3V_2(PO_4)_2F_3$ for $x = 0$ (**Figure 2a**) and $Na_3V_2O_2(PO_4)_2F$ for $x = 1$ (**Figure 2b**), whereas intermediate compounds ($0 < x < 1$) would be V^{3+}/V^{4+} mixed valence phases (**Figure 3**). The existence of a mixed valence family of compounds between $Na_3V_2(PO_4)_2F_3$ and $Na_3V_2O_2(PO_4)_2F$ phases was also reported by Park *et al.* [41] and more recently, by Qi *et al.* [42].

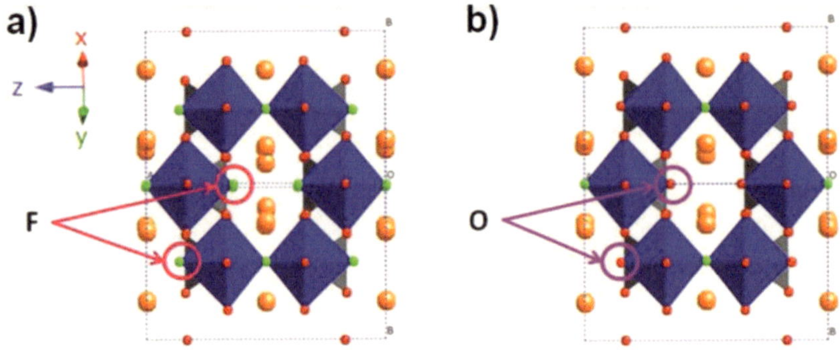

Figure 2. Crystallographic structure of a) $Na_3V_2(PO_4)_2F_3$ and b) $Na_3V_2O_2(PO_4)_2F$. Oxygen in red and fluorine in green. Vanadium octahedron is represented in blue and phosphate tetrahedron in grey.

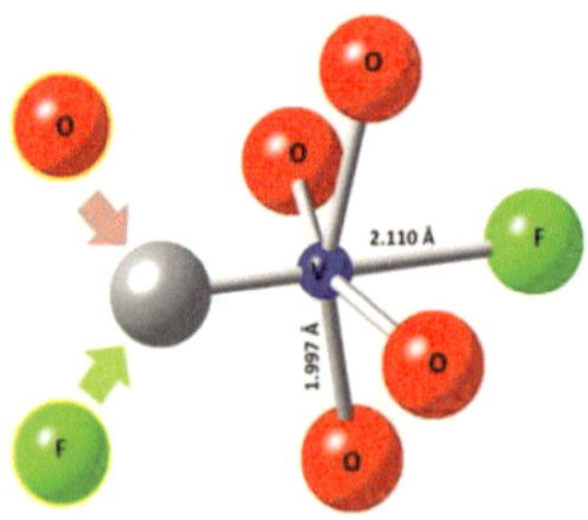

Figure 3. Vanadium octahedron for mixed valent $Na_3V_2O_{2x}(PO_4)_2F_{3-2x}$ ($0 < x < 1$) compounds. Vanadium is represented in blue, oxygen in red and fluorine in green.

In order to understand the structure and properties of this family of compounds, a series of sodium–vanadium fluorophosphate samples were hydrothermally prepared, varying the type and amount of carbon used as reductive agent during the synthesis [40]. This way, the possible influence of that carbon on both the final properties of the sample and the attainment of different compositions in the family of general formula $Na_3V_2O_{2x}(PO_4)_2F_{3-2x}$ ($0 < x < 1$) was analysed. Intermediate amounts of carbon present in the $Na_3V_2O_{2x}(PO_4)_2F_{3-2x}$ final product (1 – 50 *wt.*% C) led to V^{3+}/V^{4+} mixed valence phases ($0 < x < 1$) while a sample containing almost only V^{4+} was obtained in a composite with less than 1 *wt.*% of carbon. The electrochemical study of these samples showed the presence of two reaction voltages at 3.6 and 4.1 V *versus* Na/Na^+ and good specific capacity values up to 100 $mAhg^{-1}$ at 1C when an electrochemical grade carbon was employed. The material containing a moderate amount of this type of carbon (~6 *wt.*%) showed the best electrochemical results. A deeper analysis of this sample with an average oxidation state of +3.8 for the vanadium, showed excellent specific capacity values from 100 to 75 $mAhg^{-1}$ for charging rates of C/20 and 5C, respectively. It is worth mentioning the high capacity retention up to 95% after 200 cycles at 1C [43]. A combined experimental and theoretical study on this family of compounds was also carried out by Park *et al.* [44,45].

http://dx.doi.org/10.5772/62317

Different samples with compositions x = 0, 0.2, 0.5, 0.8 and 1 were synthesised by solid state reactions. The obtained compounds were characterised by the use of different techniques such as X-ray Diffraction (XRD), Transmission Electron Microscopy (TEM), Nuclear Magnetic Resonance (NMR), etc. The electrochemical study showed specific capacity values of approximately 100 $mAhg^{-1}$ only in the case of the V^{4+} composition. The specific capacity value was maintained for all the compounds during 150 cycles at C/2. Recently and in order to achieve different compositions of the family of sodium–vanadium fluorophosphates a new synthesis based on the solvothermal method at low-temperature (60-120°C) has been employed [42]. The size of the particles obtained with this method was in the range of 20–50 nanometers. The electrochemical analysis of these compounds showed a rate capability of 73 $mAhg^{-1}$ at 10C rate and long cycle stability over 1200 cycles at 2C.

Regarding the V^{4+} phase, $Na_3V_2O_2(PO_4)_2F$, the first electrochemical results were presented by Sauvage *et al.* getting a capacity of 87 $mAhg^{-1}$ at C/100 through a two voltage-step behaviour at 3.6 and 4.0 V vs. Na/Na^+. They found that the obtained electrochemical curve was almost identical to the ones reported for $Na_3V_2(PO_4)_2F_3$ and $NaVPO_4F$ phases giving rise to doubt about the real existence of this latter one in light of lack reliable information about its structure [27]. Some years after, a novel single-step hydrothermal treatment was discovered to obtain the $Na_3V_2O_2(PO_4)_2F$ phase [46]. The absence of *in-situ* carbon in the synthesised product made it necessary to employ an *ex-situ* carbon coating in order to improve the conductivity of the material and hence its electrochemical performance. However, control of the thermal treatment process resulted determinant to avoid the degradation of the material and the appearance of secondary phases such as NASICON $Na_3V_2(PO_4)_3$. The carbon coated material prepared by impregnation and a flash thermal treatment, exhibited good rate capabilities between 100 and 65 $mAhg^{-1}$ for rates of C/20 and 1C respectively and good cycling stability (**Figure 4**).

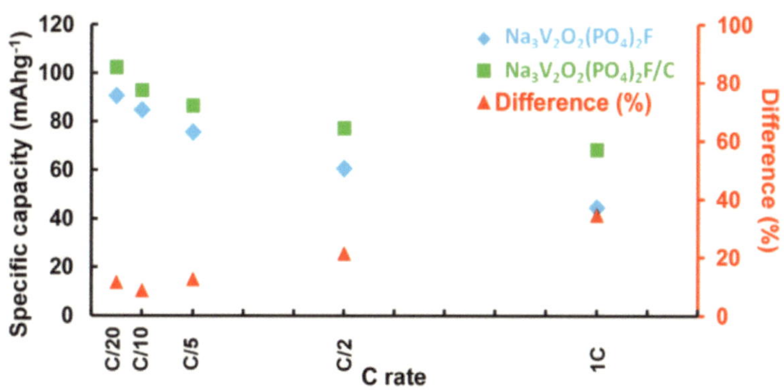

Figure 4. Comparison of discharge specific capacity values between raw $Na_3V_2O_2(PO_4)_2F$ and carbon coated flash thermally treated $Na_3V_2O_2(PO_4)_2F/C$ samples.

Additional works showed that the reversible capacity of $Na_3V_2O_2(PO_4)_2F$/graphene electrodes exceed 100 $mAhg^{-1}$ after 200 cycles at a C/20 current rate [37]. Apart from that, a theoretical and experimental study on a solvothermally obtained $Na_3V_2O_2(PO_4)_2F$ phase was presented were structural and electrochemical characterisation of this material was carried out [18].

Moreover, computational studies revealed the possibility of extracting the third sodium of the mentioned phase at 5.3 V *versus* sodium. However, this would be experimentally complicated due to the overcoming of the stability window of the electrolyte (4.8 V). However, a hypothetical new vanadium fluorophosphate phase, $Na_3V_2Cl_2(PO_4)_2F$, where the oxygen is substituted by chlorine , was proposed from the theoretical study. This latter compound would provide 188 $mAh{\cdot}g^{-1}$ theoretical specific capacity and 758 $Wh{\cdot}kg^{-1}$ energy density. Finally, electrochemical results of RuO_2-coated $Na_3V_2O_2(PO_4)_2F$ nanowires were presented were a reversible capacity of 120 $mAh{\cdot}g^{-1}$ was achieved at 1C rate during more than 1000 cycles [47].

Concerning the V^{+3} phase, $Na_3V_2(PO_4)_2F_3$, the electrochemical mechanism of the reversible Na extraction was described by Chihara *et al.* [48] and Shakoor *et al.* [38]. Both groups reported that the electrochemical reaction occurs through a single-phase reaction with negligible volume variation (~2%). More recently, the ion migration mechanism and diffusion capability of the $Na_3V_2(PO_4)_2F_3$ phase have been explored [49,50]. However, the presented results could correspond to any of the sodium–vanadium fluorophosphates included in the general formula $Na_3V_2O_{2x}(PO_4)_2F_{3-2x}$. Moreover, one of the three discharge *plateaux* that they achieve in the phase (the one at 3.4 V) could correspond to the non fluorinated NASICON $Na_3V_2(PO_4)_3$ phase as recently reported[46]. On the other hand, an *ex situ* high resolution ^{23}Na and ^{31}P solid state NMR study on this $Na_3V_2(PO_4)_2F_3$ phase was reported [51]. It is mentioned that ^{23}Na NMR spectrum of pristine $Na_3V_2(PO_4)_2F_3$ phase exhibits two major resonances at 92 and 146 ppm, which are assigned to partially filled Na(2) site and fully occupied Na(1) site, respectively. However, in recent works concerning ssNMR of the mixed valent sodium–vanadium fluorophosphate phase the peaks found in the spectra were assigned to the presence of different vanadium environments instead of the occupancy of the Na cations [44,52]. Similar results were found in the case of the $Na_3V_2O_2(PO_4)_2F$ phase containing only V^{4+} . The results of the ^{23}Na ssNMR spectrum obtained in this phase show the presence of a unique peak with an occupancy factor of the two Na cations of 1 for Na(1) site and ½ for Na(2) site (see **Figure 5**) [53].

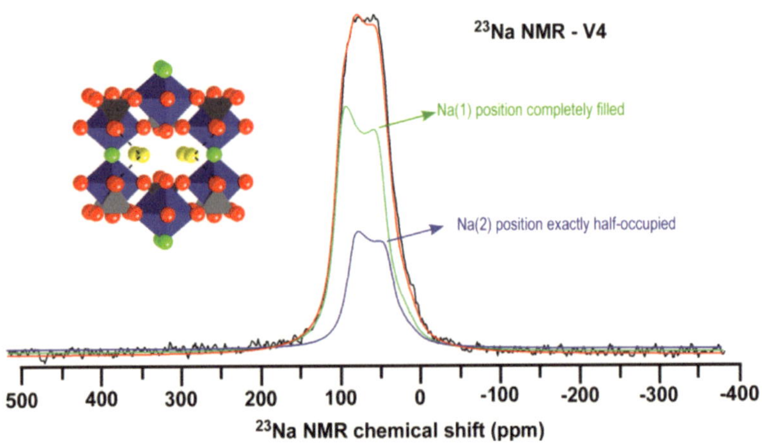

Figure 5. ^{23}Na ssNMR spectrum of $Na_3V_2O_2(PO_4)_2F$ sample (black line). Fittings for Na(1) and Na(2) are shown in green and blue, respectively.

http://dx.doi.org/10.5772/62317

The crystal structure of the $Na_3V_2(PO_4)_2F_3$ phase at high temperature (400 K) obtained from high resolution synchrotron radiation measurements in *operando* revealed the existence of a combined solid solution region and four intermediate phases during the charge process [33,34]. In addition, a new space group (*Amam)* was achieved for the $Na_3V_2(PO_4)_2F_3$ phase . On the other hand, the electrochemical results obtained in a full cell using $Na_3V_2(PO_4)_2F_3$ as cathode, *hard carbon* as anode and an optimised electrolyte showed a capacity of 97 $mAh{\cdot}g^{-1}$ after more than 120 cycles at C/5 and 70 $mAh{\cdot}g^{-1}$ at 5C rate. [39].

Concerning *ex-situ* structural study of the aforementioned phases, a detailed study was recently presented [44,45] to determine the sodium extraction/insertion mechanism of the mixed valent fluorophosphate materials. Examination of the *ex-situ* XRD patterns of $V^{3.8+}$ $Na_3V_2O_{1.6}(PO_4)_2F_{1.4}$ compound revealed the coexistence of a two-phase region in the first half of the charge indicating the presence of a stable intermediate phase between the two *plateaux* at 3.6 and 4.1 V *versus* Na/Na^+. It was also reported that the V^{3+} present in the $V^{3.8+}$ $Na_3V_2O_{1.6}(PO_4)_2F_{1.4}$ phase creates a local environment similar to the one presented by $Na_3V_2(PO_4)_2F_3$ phase [38] leading to a high-voltage V^{3+}/V^{4+} redox reaction. It is assumed that 2.4 electrons are supplied from the $V^{3.8+}$ / V^{5+} redox couple leading to a theoretical specific capacity of 155.6 $mAhg^{-1}$. On the other hand, X-Ray Absorption Near Edge Spectroscopy (XANES) studies revealed that an oxidation process happens while charging and that the reverse process occurs when discharging but no evolution of the value of the vanadium oxidation state was determined.

Concerning $Na_3V_2O_2(PO_4)_2F$ phase, different space group symmetries have been described in the literature to define it [28,29]. In terms of sodium insertion/extraction, the *I4/mmm* is shown to reversibly cycle 1.12 Na ions per formula unit ($Na_3V_2O_2(PO_4)_2F$) [27] while the *$P4_2/nmn$* polymorph with a carbon-coated synthetic procedure, reversibly cycles 1.56 Na ions per formula unit [46].

Finally, different time-resolved *in-situ* synchrotron X-ray powder diffraction studies have been carried out in order to analyse reaction mechanism evolution, lattice parameters ,and sodium extraction of $V^{3.8+}$($Na_3V_2O_{1.6}(PO_4)_2F_{1.4}$), V^{4+} $Na_3V_2O_2(PO_4)_2F$ and V^{+3} $Na_3V_2(PO_4)_2F_3$ materials [34,54,55].

The electrochemical behaviour of the two electrode materials studied *versus* Na/Na^+ showed similar electrochemical reaction mechanism evolution where a simultaneous solid solution and two-phase region were observed. However, the maximum sodium amount that can be extracted when a potentiostatic step is applied (2 hours at 4.3 V) is higher for $V^{3.8+}$ ($Na_3V_2O_{1.6}(PO_4)_2F_{1.4}$) sample than for the V^{4+} $Na_3V_2O_2(PO_4)_2F$ one, 2.23 *versus* 1.75, respectively. This fact can be due to several reasons, such as better intrinsic conductivity, the presence of *in-situ* carbon in the mixed valent sample or to the more rigid structure of the only V^{4+} containing sample. Nevertheless, both materials present high reversibility and flexibility during charge/discharge process. The effects of overcharging up to 4.8 V and cycling-related degradation of the same two compounds, $V^{3.8+}$($Na_3V_2O_{1.6}(PO_4)_2F_{1.4}$) and V^{4+} $Na_3V_2O_2(PO_4)_2F$ [56] have been recently reported. The $V^{3.8+}$($Na_3V_2O_{1.6}(PO_4)_2F_{1.4}$) phase presents less stress than V^{4+} phase due to the more disordered structure and the presence of *in situ* carbon. Thus, this phase is able to reach the maximum theoretical Na^+ extraction limit when charging up to 4.8 V while in the case of V^{4+} $Na_3V_2O_2(PO_4)_2F$ phase this limit is established at the usual 4.3 V cut-off voltage. Moreover, mid-term cycling shows the formation of secondary products due to the degradation of the materials in both cases.

To finish with, the structural evolution of V^{3+} $Na_3V_2(PO_4)_2F_3$ phase has recently been analysed by using high angular and intensity resolution synchrotron radiation [34]. The two voltage domains (at 3.7 and 4.2 V *versus* Na/Na^+) have been studied separately. In the case of the low voltage *plateau* (3.7 V) three biphasic reactions have been observed while in the higher voltage region (4.2 V) a small solid solution zone together with a biphasic zone is shown (**Figure 6**). Unexpectedly, the structure of the $NaV_2(PO_4)_2F_3$ end member is different from the previous ones revealing a different space group (*Cmc2₁*) and two different vanadium environments (V^{3+} and V^{5+}).

4. Conclusions

Polyanions $(XO_4)^{n-}$ based on transition metals (M) are one of the most promising candidates for high-voltage Na-ion batteries. Among them, sulphate and phosphate based framework materials stand out. Regarding phosphate-based materials, and as it has been previously shown in this chapter, there has been some controversy about the structural and electrochemical features of sodium–vanadium fluorophosphates. This family of materials show very high

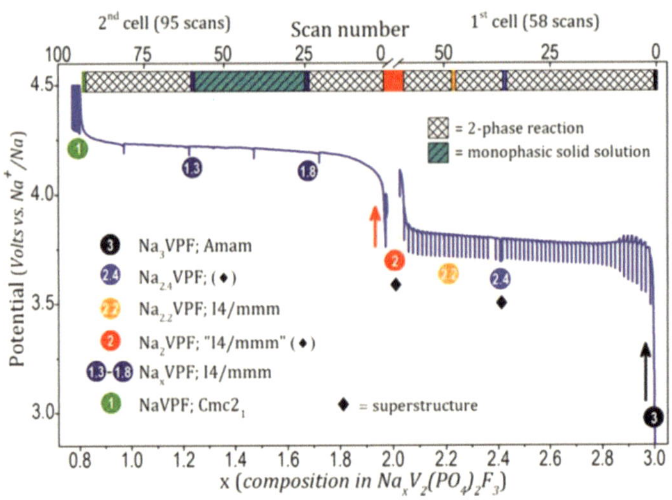

Figure 6. Potential–composition electrochemical curves obtained upon Na+ extraction from Na3VPF. The first cell was cycled (GITT) between the compositions Na_3VPF and Na_2VPF while recording 58 5 min diffraction scans; the second one (galvanostatic cycling) between the compositions Na_2VPF and Na_1VPF while recording 95 5 min scans. Significant single-phase compositions are highlighted by coloured circles. For the first cell they are obtained by the electrochemistry. Cycling of the second cell resulted in $\Delta x > 1$ because of electrolyte oxidation, hence in this case reported compositions are those obtained from Rietveld refinement. Top line describes the nature of the observed deintercalation mechanism. Reproduced from reference [34].

http://dx.doi.org/10.5772/62317

energy density due to the high voltage provided by the adding inductive effect of fluorine. Moreover, these materials have demonstrated to be very stable and present outstanding electrochemical properties. V^{+3} $Na_3V_2(PO_4)_2F_3$, $V^{+3.8}$ $Na_3V_2O_{1.6}(PO_4)_2F_{1.4}$ and V^{+4} $Na_3V_2O_2(PO_4)_2F$ phases included in the family of compounds $Na_3V_2O_{2x}(PO_4)_2F_{3-2x}$ have been structurally and electrochemically deeply analysed. Despite the similarity of all these materials, up to date studies show that the electrochemical behaviour of these electrode materials is clearly dependant not only on the sodium extraction/insertion mechanism, the occupancy and distribution of sodium and the electrochemical cycling history but also on the synthesis process employed for the obtaining of the starting material that determines its properties.

5. Future perspectives for sodium–vanadium fluorophosphates as cathodes for Na-ion batteries

As it has been presented, sodium–vanadium fluorophosphates are more than promising cathodic materials for near future commercial Na-ion batteries. However, the appearance of real commercial Na-ion batteries involves the development of the whole battery, that includes advancing in finding the appropriate components for a Na-ion cell based on a sodium–vanadium fluorophosphate cathode. In the first place, an anode matching this cathode must be chosen. This anode material should present relatively low operating voltage in order to provide the battery with the maximum available voltage. In the second place, the specific capacity of this anode and its rate capability should be in balance with the cathode performance in order to make an easier mass balance to prepare the cells. These two conditions will be necessary to assure that the battery, and thus its components, present the best possible performance in terms of battery voltage and power.

On the other hand, a special effort must be done to search for a long-lasting and high-voltage working electrolyte for these systems. Since the recovery of Na-ion battery research, great efforts have been directed to the search of new electrode materials whereas studies dealing with the electrolyte are much scarcer. One of the reasons for this trend can be that it has been shown that the SEI formed on carbonaceous electrodes is markedly different for sodium- and lithium-based electrolytes even using the same solvent [57,58]. Non-aqueous Na electrolytes presently used are mainly based on $NaClO_4$ or $NaPF_6$ salts in propylene carbonate or other solvents and mixtures such as ethylene carbonate: dimethyl carbonate. In general, a good electrolyte should exhibit: i) good ionic conductivity, ii) a large electrochemical window (i.e., high and low onset potential for electrolyte decomposition through oxidation and reduction at high and low voltages, respectively), iii) no reactivity towards the cell components, iv) thermal stability (i.e. melting point and boiling point lower and higher than the standard temperatures for the cell utilization, respectively). Finally, it should be intrinsically safe, have as low toxicity as possible and meet cost requirements for the targeted applications. All these features are intrinsically dependent on the nature of the salt and the solvent(s) and the possible use of additives [59].

Another argument pointed out as a drawback for Na-ion batteries based on sodium–vanadium fluorophosphates is the use of a vanadium compound in them. It is said that vanadium is not a green element that needs careful handling after battery life cycle, but the same can be said from the electrolytes used in both sodium- and lithium-based systems. In any case, a battery will always need to be collected to be treated as a residue after its cycle life, as it is nowadays done. Apart from that, it is also commented that vanadium cost is another drawback for the development of these batteries. But it must be recalled that there exist commercial vanadium-rich energy storage systems, such as redox flow batteries, that comprise the use of high amounts of vanadium for their two electrodes [60]. Thus, a careful cost analysis should be done to confirm or discard this statement once the appropriate anode and electrolyte materials are chosen and a viable cell prototype is proposed.

Finally, and now focusing our attention on the commercial development of only sodium–vanadium fluorophosphates compounds, a last challenge that must be overcome is to develop a synthetic method that allows an easy and economical mass production of these materials. For this purpose, different parameters will have to be taken into account: the different possible synthetic methods (hydrothermal, microwave or ceramic processing); the different possible vanadium starting reactants; and the yields obtained by crossing these latter two parameters (**Figure 7**).

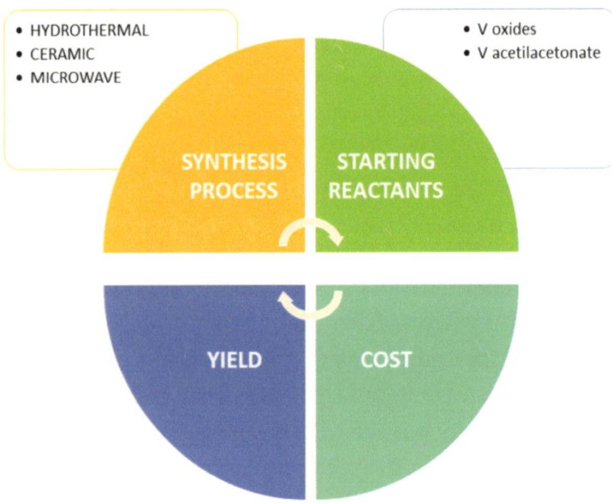

Figure 7. Main parameters that must be adjusted to get a commercially viable production method for sodium–vanadium fluorophosphates.

To sum up, sodium–vanadium fluorophosphates family is one of the main possible future cathodic materials for high-voltage Na-ion batteries. Their excellent electrochemical properties and performance have been perfectly well demonstrated by different research groups, as it has been told in this chapter. Now, the great defy for these materials lays on the industrial pace, that is, in the assembly and building of a real Na-ion full cell based on these materials and the development of a commercially viable process to produce these compounds in great amounts.

http://dx.doi.org/10.5772/62317

Acknowledgements

This work was financially supported by the Ministerio de Economía y Competitividad (MAT2013-41128-R) and the Gobierno Vasco/Eusko Jaurlaritza (IT570-13). University of the Basque Country (UPV/EHU) is acknowledged for funding under project UFI11/15.

Author details

Paula Serras[1,2*], Verónica Palomares[1] and Teófilo Rojo[1,3]

*Address all correspondence to: paula_serras@yahoo.es

1 Inorganic Chemistry Department, University of the Basque Country UPV/EHU, Leioa, Bizkaia, Spain

2 Nuclear Engineering and Fluids Mechanics Department, Faculty of Engineering, University of the Basque Country UPV/EHU, Bilbao, Bizkaia, Spain

3 CIC Energigune, Alava Technology Park, Miñano, Álava, Spain

References

[1] Pavley F, Nunez F. California Assembly Bill No. 32-Global Warming Solutions Act of 2006. 2006.

[2] CARPS, California Codes: Public Utilities Code. Section 399.11-399-31. California Renewables Portfolio Standard Program. 2009.

[3] Ellis B L, Makahnouk W R M, Makimura Y, Toghill K, Nazar L F. A multifunctional 3.5 V iron-based phosphate cathode for rechargeable batteries. Nature Materials. 2007: 6: 749–753. DOI: 10.1038/nmat2007

[4] Holleman A F, Wiberg N. Lehrbuch der Anorganischen Chemie. 102nd ed. Berlin: De Gruyter; 2007. ISBN: 978-3-11-017770-1

[5] Palomares V, Serras P, Villaluenga I, Hueso K B, Carretero-González J, Rojo T. Na-ion batteries, recent advances and present challenges to become low cost energy storage systems. Energy & Environmental Science. 2012: 5:5884–5901. DOI: 10.1039/c2ee02781j

[6] Ong S.P, Chevrier V.L, Hautier G, JainA, Moore C, Kim S, Ma X, Ceder G. Voltage, stability and diffusion barrier differences between sodium-ion and lithium-ion intercalation materials. Energy & Environmental Science. 2011: 4: 3680–3688. DOI: 0.1039/c1ee01782a.

[7] Sawicki M, Shaw L L. Advances and challenges of sodium ion batteries as post lithium ion batteries. RSC Advances. 2015: 5: 53129–53154. DOI: 10.1039/c5ra08321d.

[8] Adelhelm P, Hartmann P, Bender C L, Busche M, Eufinger C, Janek J. From lithium to sodium: cell chemistry of room temperature sodium–air and sodium–sulfur batteries. Beilstein Journal of Nanotechnology. 2015: 6: 1016–1055. DOI: 10.3762/bjnano.6.105

[9] Masquelier C, Croguennec L. Polyanionic (phosphates, silicates, sulfates) frameworks as electrode materials for rechargeable Li (or Na) batteries. Chemical Reviews, 2013: 113: 6552–6591. DOI: 10.1021/cr3001862.

[10] Manthiram A, Goodenough J.B. Lithium insertion into $Fe_2(SO_4)_3$ frameworks. Journal of Power Sources, 1989: 26: 403–408. DOI: 10.1016/0378-7753(89)80153-3.

[11] Barpanda P. Sulfate Chemistry for high-voltage insertion materials: Synthetic, structural and electrochemical insights. Israel Journal of Chemistry. 2015: 55: 537–557. DOI: 10.1002/ijch.201400157

[12] Barpanda P, Oyama G, Nishimura S.I, Chung S.C., Yamada A. A 3.8-V earth-abundant sodium battery electrode. Nature Communications. 2014:5: 1–8. DOI: 10.1038/ncomms5358.

[13] Sawicki M, Shaw L.L. Advances and challenges of sodium ion batteries as post lithium ion batteries. RSC Advances. 2015:5: 53129–53154. DOI: 10.1039/c5ra08321d.

[14] Yamada A, Chung S.C, Hinokuma K. J. Optimised $LiFePO_4$ for lithium battery cathodes. Electrochemical Society. 2001: 148: A224–A229. DOI: 10.1149/1.1348257.

[15] Lee K.T, Ramesh T.N, Nan F, Botton G, Nazar L.F. Topochemical synthesis of sodium metal phosphate olivines for sodium—ion batteries. Chemistry of Materials. 2011: 23: 3493–3600. DOI: 10.1021/cm200450y.

[16] Boyadzhieva T, Koleva V, Zhecheva E, Nihtianova D, Mihaylovc L, Stoyanova R. Competitive lithium and sodium intercalation into sodium manganese phospho-olivine $NaMnPO_4$ covered with carbon black. RSC Advances. 2015: 5: 87694–87705. DOI: 10.1039/c5ra17299c.

[17] Nose M, Nakayama H, Nobuhara K, Yamaguchi H, Nakanishi S, Iba H. $Na_4Co_3(PO_4)_2P_2O_7$: A novel storage material for sodium-ion batteries. Journal of Power Sources. 2013: 234: 175–179. DOI: 10.1016/j.jpowsour.2013.01.162.

[18] Xu M, Xiao P, Stauffer S, Song J, Henkelman G, Goodenough J.B. Theoretical and experimental study of vanadium-based fluorophosphate cathodes for rechargeable batteries. Chemistry of Materials. 2014: 26: 3089–3097. DOI: 10.1021/cm500106w.

[19] Ellis B.L, Makahnouk W.R.M, Rowan-Weetaluktuk W.N, Ryan D.H, Nazar L.F. Crystal structure and electrochemical properties of A_2MPO_4F fluorophosphates (A = Na, Li; M = Fe, Mn, Co, Ni). Chemistry of Materials. 2010: 22: 1059–1070. DOI: 10.1021/cm902023h.

[20] Yakubovich O.V, Karimova O.V, Mel'nikov O.K. The mixed anionic framework in the structure of $Na_2MnF(PO_4)$. Acta Crystallographica Section C. 1997: 53: 395–397. DOI: 10.1107/S0108270196014102.

[21] Langrock A, Xu Y, Liu Y, Ehrman S, Manivannan A, Wang C. Carbon coated hollow Na_2FePO_4F spheres for Na–ion battery cathodes. Journal of Power Sources. 2013: 223: 62–67. DOI: 10.1016/j.jpowsour.2012.09.059.

[22] Barker J, Saidi M.Y, Swoyer J.L. A sodium-ion cell based on the fluorophosphate compound $NaVPO_4$ F. Electrochemical and Solid-State Letters. 2003: 6: A1–A4. DOI: 10.1149/1.1523691.

[23] Zhuo H, Wang X, Tang A, Liu Z, Gamboa S, Sebastian P.J. The preparation of $NaV_{1-x}Cr_xPO_4F$ cathode materials for sodium-ion battery. Journal of Power Sources. 2006: 160: 698–703. DOI: 10.1016/j.jpowsour.2005.12.079.

[24] Liu Z, Wang X, Wang Y, Tang A, Yang S, He L. Preparation of $NaV_{1-x}Al_xPO_4F$ cathode materials for application of sodium-ion battery. Transactions of Nonferrous Metals Society of China. 2008: 18: 346–350. DOI: 10.1016/S1003-6326(08)60060-6.

[25] Lu Y, Zhang S, Li Y, Xue L, Xu G, Zhang X. Preparation and characterisation of carbon-coated $NaVPO_4F$ as cathode material for rechargeable sodium-ion batteries. Journal of Power Sources. 2013: 247: 770–777. DOI: 10.1016/j.jpowsour.2013.09.018.

[26] Xu M, Cheng C.-J, Sun Q.-Q, Bao S.-J, Niu Y.-B, He H, Li Y, Song J. A 3D porous interconnected $NaVPO_4F/C$ network: preparation and performance for Na–ion batteries. RSC Advances. 2015;5: 40065–40069. DOI: 10.1039/c5ra05161d.

[27] Sauvage F, Quarez E, Tarascon J, Baudrin E. Crystal structure and electrochemical properties vs. Na^+ of the sodium fluorophosphate $Na_{1.5}VOPO_4F_{0.5.}$ Solid State Sciences. 2006; 8: 1215–1221. DOI: 10.1016/j.solidstatesciences.2006.05.009.

[28] Massa W, Yakubovich O, Dimitrova O. Crystal structure of a new sodium vanadyl(IV) fluoride phosphate $Na_3V_2O_2F[PO_4]_2$. Solid State Sciences. 2002;4: 495–501. DOI: 10.1016/S1293-2558(02)01283-9.

[29] Tsirlin A, Nath R, Abakumov A, Furukawa Y, Johnston D, Hemmida M, Krug von Nidda H.-A, Loidl A, Geibel C, Rosner H. Phase separation and frustrated square lattice magnetism of $Na_{1.5}VOPO_4F_{0.5}$. Physical Review B. 2011;84: 014429/1-014429/16. DOI: 10.1103/PhysRevB.84.014429.

[30] Le Meins J.-M, Crosnier-Lopez M.-P, Hemon-Ribaud A, Courbion G. Phase Transitions in the $Na_3M_2(PO_4)_2F_3$ Family (M=Al^{3+}, V^{3+}, Cr^{3+}, Fe^{3+}, Ga^{3+}): Synthesis, Thermal, Structural, and Magnetic Studies. Journal of Solid State Chemistry. 1999;148: 260–277. DOI: 10.1006/jssc.1999.8447.

[31] Barker J, Gover R.K.B, Burns P, Bryan A.J. Hybrid-Ion: A lithium–ion cell based on a sodium insertion material. Electrochemical Solid-State Letters. 2006;9: A190-A192. DOI: 10.1149/1.2168288.

[32] Jiang T, Chen G, Li A, Wang C, Wei Y. Sol–gel preparation and electrochemical properties of $Na_3V_2(PO_4)_2F_3/C$ composite cathode material for lithium ion batteries. Journal of Alloys and Compounds. 2009;478: 604–607. DOI: 10.1016/j.jallcom.2008.11.147.

[33] Bianchini M, Brisset N, Fauth F, Weill F, Elkaim E, Suard E, Masquelier C, Croguennec L. $Na_3V_2(PO_4)_2F_3$ revisited: A high resolution diffraction study. Chemistry of Materials. 2014;26:4238–4247. DOI: 10.1021/cm501644g.

[34] Bianchini M, Fauth F, Brisset N, Weill F, Suard E, Masquelier C, Croguennec L. Comprehensive investigation of the $Na_3V_2(PO_4)_2F_3$ - $NaV_2(PO_4)_2F_3$ system by operand high resolution synchrotron X-ray diffraction. Chemistry of Materials. 2015;27:3009–3020. DOI: 10.1021/acs.chemmater.5b00361.

[35] Ateba J.-M, Masquelier C, Suard E, Croguennec L. Synthesis and crystallographic study of homeotypic $LiVPO_4F$ and $LiVPO_4O$. Chemistry of Materials. 2012;24: 1223–1234. DOI: 10.1021/cm3003996.

[36] Ruan Y.-L, Wang K, Song S.-D, Han X, Cheng B.W. Graphene modified sodium vanadium fluorophosphate as high voltage cathode material for sodium ion batteries. Electrochimica Acta. 2015;160:330-336. DOI: 10.1016/j.electacta.2015.01.186

[37] Xu M, Wang L, Zhao X, Song J, Xie H, Lu Y, Goodenough J.B. $Na_3V_2O_2(PO_4)_2F$/graphene sandwich structure for high-performance cathode of a sodium-ion battery. Physical Chemistry Chemical Physics. 2013;15: 13032–13037. DOI: 10.1039/C3CP52408F.

[38] Shakoor R.A, Seo D.-H, Kim H, Park Y.U, Kim J, Kim S.-W, Gwon H, Lee S, Kang K. A combined first principles and experimental study on $Na_3V_2(PO_4)_2F_3$ for rechargeable Na batteries. Journal of Materials Chemistry. 2012;22: 20535–20541. DOI: 10.1039/C2JM33862A.

[39] Ponrouch A, Dedryvère R, Monti D, Demet A.E, Ateba J.M, Croguennec L, Masquelier C, Johansson P, Palacín M.R. Towards high energy density sodium ion batteries through electrolyte optimisation. Energy and Environmental Science. 2013;6: 2361–2369. DOI: 10.1039/C3EE41379A.

[40] Serras P, Palomares V, Goñi A, Gil de Muro I, Kubiak P, Lezama L, Rojo T. High voltage cathode materials for Na–ion batteries of general formula $Na_3V_2O_{2x}(PO_4)_2F_{3-2x}$. Journal of Materials Chemistry. 2012;22: 22301–22308. DOI: 10.1039/C2JM35293A.

[41] Park Y.U, Seo D.H, Kim B, Hon K.P, Kim H, Lee S, Shakoor R.A, Miyasaka K, Tarascon J.M, Kang K. Tailoring a fluorophosphate as a novel 4 V cathode for lithium-ion batteries. Scientific Reports. 2012;2: 704. DOI: 10.1038/srep00704.

[42] Qi Y, Mu L, Zhao J, Hu Y.-S, Liu H, Dai S. Superior Na-storage performance of low-temperature-synthesized $Na_3(VO_{1-x}PO_4)2F_{1+2x}$ $(0 \leq x \leq 1)$ nanoparticles for Na–ion batteries. Angewandte Chemie International Edition. 2015;54: 9911–9916. DOI: 10.1002/anie.201503188.

[43] Serras P, Palomares V, Goñi A, Kubiak P, Rojo T. Electrochemical performance of mixed valence $Na_3V_2O_{2x}(PO_4)_2F_{3-2x}$/C as cathode for sodium-ion batteries. Journal of Power Sources. 2013;241: 56–60. DOI: 10.1016/j.jpowsour.2013.04.094.

[44] Park Y.-U, Seo D. H, Kwon H.-S, Kim B, Kim J, Kim H, Kim I, Yoo H.-I, Kang K. A new high-energy cathode for a Na–ion battery with ultrahigh stability. Journal of the American Chemical Society. 2013;135: 13870–13878. DOI: 10.1021/ja406016j.

[45] Park Y. U, Seo D. H, Kim B, Kim J, Lee S, Kim B, Kang K. A family of high-performance cathode materials for Na–ion batteries, $Na_3(VO_{1-x}PO_4)_2$ F_{1+2x} ($0 \leq x \leq 1$): combined first-principles and experimental study. Advanced Functional Materials. 2014;24: 4603–4614. DOI: 10.1002/adfm.201400561.

[46] Serras P, Palomares V, Kubiak P, Lezama L, Rojo T. Enhanced electrochemical performance of vanadyl (IV) $Na_3(VO)_2(PO_4)_2F$ by ex-situ carbon coating. Electrochemistry Communications. 2013;34: 344–347. DOI: 10.1016/j.elecom.2013.07.010.

[47] Peng M, Li B, Yan H, Zhang D, Wang X, Xia D, Guo G. Ruthenium-oxide-coated sodium vanadium fluorophosphate nanowires as high power cathode materials for sodium ion batteries. Angewandte Chemie International Edition. 2015;54: 6452–6456. DOI: 10.1002/anie.201411917.

[48] Chihara K, Kitajou A, Gocheva I.D, Okada S, Yamaki J.-I. Cathode properties of $Na_3M_2(PO_4)_2F_3$ [M = Ti, Fe, V] for sodium ion batteries. J. Power Sources. 2013;227: 80–85. DOI: 10.1016/j.jpowsour.2012.10.034.

[49] Song W, Ji X, Wu Z, Yang Y, Zhou Z, Li F, Chen Q, Banks C.E. Exploration of ion migration mechanism and diffusion capability for $Na_3V_2(PO_4)_2F_3$ cathode utilized in rechargeable sodium ion batteries. Journal of Power Sources. 2014;256:258–263. DOI: 10.1016/j.jpowsour.2014.01.025.

[50] Song W, Ji X, Wu Z, Zhu Y, Li F, Yao Y, Banks C.E. Multifunctional dual $Na_3V_2(PO_4)_2F_3$ cathode for both lithium-ion and sodium-ion batteries. RSC Advances. 2014;4:11375–11383. DOI: 10.1039/c3ra47878e.

[51] Liu Z, Hu Y.-Y, Dunstan M.T, Huo H, Hao X, Zou H, Zhong G, Yang Y, Grey C.P. Local structure and dynamics in the Na ion battery positive electrode material $Na_3V_2(PO_4)_2F_3$. Chemistry of Materials. 2014;26:2513–2521. DOI: 10.1021/cm403728w.

[52] Serras P, Palomares V, Alonso J, Sharma N, Lopez del Amo J.M, Kubiak P, Fdez-Gubieda M.L, Rojo T. Electrochemical Na extraction/insertion of $Na_3V_2O_{2x}(PO_4)_2F_{3-2x}$. Chemistry of Materials. 2013;25:4917–4925. DOI: 10.1021/cm403679b.

[53] Serras Malillos P. High voltage cathodes for Na–ion batteries: $Na_3V_2O_{2x}(PO_4)_2F_{3-2x}$ system [thesis]. Leioa: University of the Basque Country; 2014.

[54] Serras P, Palomares V, Rojo T, Brand H.E.A, Sharma N. Structural evolution of high energy density V^{3+}/V^{4+} mixed valent $Na_3V_2O_{2x}(PO_4)_2F_{3-2x}$ (x = 0.8) sodium vanadium

fluorophosphate using *in situ* synchrotron X-ray powder diffraction. Journal of Materials Chemistry A. 2014;2:7766–7779. DOI: 10.1039/c4ta00773e.

[55] Sharma N, Serras P, Palomares V, Brand H.E.A, Alonso J, Kubiak P, Fdez-Gubieda M.L, Rojo T. Sodium distribution and reaction mechanisms of a $Na_3V_2O_2(PO_4)_2F$ electrode during use in a sodium-ion battery. Chemistry of Materials. 2014;26:3391–3402. DOI: 10.1021/cm5005104.

[56] Palomares V, Serras P, Brand H.E.A, Rojo T, Sharma N. Structural evolution of mixed valent (V^{3+}/V^{4+}) and V^{4+} sodium vanadium fluorophosphates as cathodes in sodium-ion batteries: comparisons, overcharging and mid-term cycling. Journal of Materials Chemistry A. 2015;3:23017–23027. DOI: 10.1039/c5ta03780h.

[57] Komaba S, Murata W, Ishikawa T, Yabuuchi N, Ozeki T, Nakayama T, Ogata A, Gotoh K, Fujiwara K. Electrochemical Na insertion and solid electrolyte interphase for hard-carbon electrodes and application to Na–Ion batteries. Advanced Functional Materials. 2011;21: 3859–3867. DOI: 10.1002/adfm.201100854.

[58] Xia X, Obrovac M N, Dahn J R. Comparison of the reactivity of Na_xC_6 and Li_xC_6 with non-aqueous solvents and electrolytes. Electrochemical and Solid State Letters. 2011;14: A130-A133. DOI.: 10.1149/1.3606364.

[59] Ponrouch A, Marchante E, Courty M, Tarascon J-M, Palacín M R. In search of an optimised electrolyte for Na–ion batteries. Energy & Environmental Science. 2012;5: 8572–8583. DOI: 10.1039/C2EE22258B.

[60] Alotto P G, Guarnieri M, Moro F. Redox flow batteries for the storage of renewable energy: A review. Renewable and Sustainable Energy Reviews. 2014;29: 325–335. DOI: 10.1016/j.rser.2013.08.001.

Chapter 9

Vanadium Pentoxide (V_2O_5) Electrode for Aqueous Energy Storage: Understand Ionic Transport using Electrochemical, X-Ray, and Computational Tools

Daniel S. Charles and Xiaowei Teng

Additional information is available at the end of the chapter

http://dx.doi.org/10.5772/62759

Abstract

In this book chapter, we have discussed the recent results on vanadium oxide-based materials for energy storage applications. Primarily, we present the new results from our own research group on V_2O_5-layered nanostructures that are made from a facile wet chemistry synthesis. By fine control of the synthetic condition, the morphology, crystallinity, and layer-to-layer distance of V_2O_5 nanostructures can be tuned. Particularly, highly disordered V_2O_5 nanolayers which have an interplanar distance up to 1.1 nm, offering a fast transport of K-ion between layers in an aqueous electrolyte and hence the high-energy storage capacity and power density. Uniqueness of our results includes materials characterization and measurements using multiple spectroscopic tools, including synchrotron X-ray pair distribution function (PDF) analyses and in situ X-ray diffraction (XRD). Combined with half-cell and button-cell electrochemical measurements, the complementary results provide insight on the ionic transport of ions between the layers of V_2O_5 nanostructure.

Keywords: Vanadium pentoxide, Electrochemical energy storage, X-ray diffraction, Pair distribution function, *in situ* XRD

1. Introduction

1.1. Electrochemical energy storage

Rising demands for energy coupled with concerns over environmental pollution and dependency on petroleum fuel have contributed to a great need for new energy conversion and storage technologies [1, 2]. According to the United States Vehicle Technologies Office (VTO) at the Office of Energy Efficiency and Renewable Energy (EERE) within the Department of Energy (DOE), transitioning to a light-duty fleet of hybrid electric vehicles and plug-in electric could reduce US foreign oil dependence by 30–60% and greenhouse gas emissions by 30–45%. Thus, two approaches, both involving electrochemical processes, are generally suggested for solving this energy dilemma for the United States: (i) Development of highly efficient electrochemical energy conversion devices, such as fuel cells, as an alternative to the internal combustion engine. Such devices can use gas or liquid fuels rather than gasoline to reduce CO_2 emissions; (ii) Development of highly efficient electrochemical energy storage (EES) devices, such as Li-, Na-, or K-ion batteries, and electrochemical capacitors, to accelerate the adoption of electrical vehicle and renewable non-carbon-emitting energy from solar and wind energy sources [3, 4].

Technology	Lithium-ion battery	Electrostatic capacitor	Electrochemical capacitor
Installation cost ($/kWh)	200–360	300–2000	9000
Installation cost ($/kW)	1000	100–300	300
Energy density (Wh/kg)	160	0.01–0.1	1–10
Power density (W/kg)	180	10^4–10^7	10–10^5
Cycle life	~10^3 cycles	>15 years	10^6 cycles

Table 1. Properties of different electrical energy storage devices.

Many EES devices, including most batteries and the majority of electrochemical capacitors,

$$MO_X + A^+ + e^- \leftrightarrow MO_XA \tag{1}$$

where MO_X is a transition metal oxide, and A^+ denotes the alkali ions (e.g., Li^+, Na^+ or K^+). Upon the insertion or de-insertion of A^+, metal (M) cations will be reduced or oxidized to balance the charge. When an EES device is charged, a voltage (V) will build up between the anode and the cathode. The energy density (E) depends on the capacitance (C) and voltage of the device, while the power density (P) depends on the voltage and equivalent internal resistance (R_S) of the device, as following equations show:

$$E = ½ C V^2 \tag{2}$$

http://dx.doi.org/10.5772/62759

$$P = V^2 / (4R_S) \tag{3}$$

To provide clean and efficient energy solutions, a new genre of EES devices that are able to fill the gap between high-energy-density battery devices and high-power-density electrostatic capacitor devices are in great demand. From Eqs. (2) and (3), it is clear that to increase energy density and power density of EES devices, one needs to:

- Synthesize nanostructured electrodes that possess tailored, high-surface-area architecture that offers more redox-active sites for electrolyte and accelerated ion conductance (to increase C and decrease R_S, and hence increase E and P);
- Modify the electrode materials by adding secondary elements which could bring new functionalities for storing multiple charges at a single site of the metal oxides nanostructure (to increase C, and hence E) and could also contribute to highly reversible charge storage for better stability;
- Discover new types of charge carrier (such as bi- and/or tri-valent cations) that may increase charge storage capacity for the same amount of ionic transport (to increase C and hence E).

1.2. Current state of knowledge of Na-ion-based aqueous EES

Current EES can be classified into aqueous and non-aqueous storage according to the chemistry of the electrolytes. Non-aqueous electrolytes can achieve wider voltage window (usually >3.0 V), and thus, non-aqueous EES usually exhibits much higher-energy density than aqueous counterparts. However, non-aqueous systems are expensive and often require the use of highly toxic and flammable solvents, which can become hazardous if used improperly, such as over charging or short-circuiting. Non-aqueous EES devices, including lithium ion batteries (LIBs), require strict preparation standards at high cost, to allow for water- and oxygen-free cell fabrication. On the other hand, aqueous EES has narrow voltage window, typically around 1.2 V, limited by oxygen and hydrogen evolution reactions of water. However, aqueous electrolytes are generally inexpensive. Water- and oxygen-fee manufacturing environment (lower than 2 ppm [part per million]) required by non-aqueous EES is unnecessary for aqueous EES, which lowers the cost of cell packing dramatically. Also, aqueous EES can use electrolytes with high salt concentration and hence have nearly two orders of magnitude higher ion conductivities compared to non-aqueous electrolytes. Aqueous lead acid batteries (LABs) are one of the most commonly used aqueous EES devices and have great advantages such as low cost and high surge current compared to LIBs. However, LABs have a short cycle life, and their components such as lead and sulphuric acid are not environmentally friendly. Other aqueous EES devices, especially rechargeable ones using earth-abundant and non-toxic electrode materials and Na- and/or K-ions as charge carriers, appear more appropriate for stationary energy storage to be used with renewable energy sources due to lower cost of raw material, longer cycling life, and better environmental friendliness compared to Li-ion and lead acid-based counterparts [5–7].

Na-ion- and K-ion-based aqueous energy storage devices have showed great potential as a new class of energy storage devices as alternatives to Li-ion storage devices, since the electrochemical potential of Na and K is almost identical to that of Li-ions. Although radii of the bare ions is in the order of $Li^+ < Na^+ < K^+$ (**Table 2**), due to the effect of hydration number (number of water molecules that migrate with each cation) of alkali metals, total radius of Na- and K-ion and their bound water molecules (Stokes radius) actually are smaller in the order of $K^+ < Na^+ < Li^+$, resulting in a higher chemical diffusion coefficient of Na- and K-ions in the aqueous electrolyte compared with Li-ions. Moreover, current mineable Li resources can be sustained for approximately only 65 years at most. In contrast, Na and K are about 1,000 times more abundant in earth crust and hence more economical. Thus, aqueous energy storage devices based on Na- and K-ions holds great promise for economical and sustainable energy storage devices with high stability, better ionic conductivity, and low cost. In this context, to find a suitable electrode materials and architecture, which guarantees a reversible intercalation and deintercalation of Na- and K-ions, while maintaining a stable electrode and electrolyte interface, becomes imperative to the design of new generation of energy storage devices.

Ionic conductivity in water ($/10^{-4}\ \Omega^{-1}\ m^2$)	Atomic radius (/pm)	Ionic radius (/pm)	Hydration number
Li 38.7	152	60	14
Na 50.1	186	95	8.4
K 73.5	227	133	5.4

Table 2. Radii and ionic conductivities of various alkali metals in water at 298 K.

1.3. Layered vanadium oxides nanomaterials for EES

Charge storage mechanisms involving intercalation and deintercalation of alkaline ions has been commonly found in ESS devices with a high-energy density, such as batteries. Although intercalation process is highly reversible and suitable for the high capacity energy storage, intercalation materials usually are not suitable for high rate performance. This is largely due to the fact that the diffusion of intercalated ions (e.g., alkaline ions) in the electrode materials is sluggish, having diffusion coefficients in the range from 10^{-13} to 10^{-17} m^2/s. To shorten the diffusion time (τ), that is the time required for alkaline ions move through the host electrode material, one has to either reduce the ionic path of alkaline ions within the electrode material (L) or increase the diffusion coefficient of alkaline ions in the solid phase (D) [8], described by the following equation:

$$\tau = L^2 / D \tag{4}$$

Since the diffusion coefficient (D) is one of the intrinsic properties of alkaline ion-host electrode material couple, the only practical way to decrease the diffusion time (τ), and hence increase the rate performance of the electrode material, is to decrease the diffusion length. In this context, diffusion time can be considered as the minimum time required by the alkaline ions to insert and become homogenously distributed within the host electrode material. Hence, a diffusion time of alkaline ions within host solid electrode is closely related to the C-rate of an EES device, which is a measure of the rate at which an EES device is discharged relative to its maximum capacity (1/n C-rate means that the discharge current will discharge the entire device in n hours).

τ (s)	C-rate	L^2 (nm^2)	L (nm)
36,000	0.1	360,000	600
3600	1	36,000	190
360	10	3600	60
36	100	360	19
10	360	100	10

Table 3. Relationship between the diffusion time (τ), C-rate, and diffusion length (L) of transport of alkaline ions within the electrode materials.

Table 3 describes relationship between the diffusion time (τ), C-rate, and diffusion length (L), assuming the diffusion coefficient of alkaline ions within metal oxide electrode materials is in the order of 10^{-17} m^2/s. Considering 36,000 s (10 h) a reasonable time to completely charge and discharge a battery device, a maximum diffusion length of 600 nm is needed. And this value appears to be easily achieved through tailoring the conventional preparation routine for the electrode materials. However, for a high-rate EES device that affords charge and discharge time around 36 s, a much shorter diffusion length of 19 nm is required, which might be too small to be prepared via conventional preparation routine, which typically involves a solid state reaction through decomposition of carbonate salts at high temperature (usually >500°C). A more likely prospective method to prepare nanoscaled electrode materials for shorter diffusion time is the synthesis of layered electrode structure. Many dioxides of transition metals consist of [MO_6] (M: Mn, Ti, V, Ir, Ru, Cr) octahedra with shared vertices and edges. Stacking of [MO_6] octahedra enables the building of two-dimensional layered structures [9–14]. Water molecules and alkaline ions such as Li^+, Na^+, or K^+ can easily move into or out of the layers without significant structural rearrangements. Particularly, large interplanar distance (usually close to 1 nm) favors the storage of alkaline cations and a higher capacitance [15].

Among various transition metal oxides, vanadium pentoxide (V_2O_5) has attracted great interest in the fields of electrochromic devices and as electrode materials for batteries and electrochemical capacitors, thanks to its low cost and ability to exist in different oxidation states from V^{2+} to V^{5+}. Layered vanadium oxides structures have been synthesized using electrodeposition, deposition under ultra-high-vacuum conditions or a top-down solution-phase synthesis [16, 17]. Particularly, solution-phase exfoliation methods that can synthesize layered nanostructures in large quantities have been well studied. High-temperature pre-treatment was usually required, and intercalators (e.g., tetrabutylammonium ions) were used to assist the separation of layers and hamper the reassembly of the bulk laminar material. Furthermore, by appropriate control of the chemical nature of the solvent, intercalators, exfoliation time, and temperature, as well as the post-treatment temperature, it is possible to tune the dimension of the interplanar distance in order to improve the electrochemical performance, that is, to make the interlayer distance large enough to be capable of accommodating the alkaline ions and water.

2. Highly crystalline V_2O_5-layered electrode material for aqueous K-ion storage

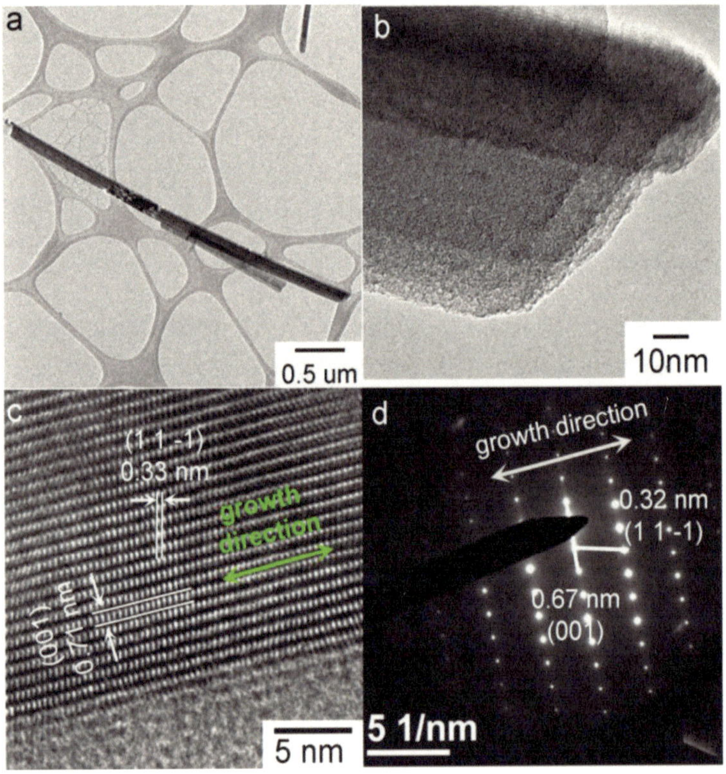

Figure 1. (a–c) TEM images of $K_{0.33}V_2O_5$-layered nanostructure after being annealed at 450°C and (d) selected-area electron diffraction of such materials.

V_2O_5-layered nanofibers can be synthesized by a facile two-step method involving solution-phase procedure followed by mild thermal treatment in the air [18]. In the first step, nearly amorphous vanadium oxides were prepared by mixing vanadyl (IV) sulfate hydrate with potassium hydroxide under vigorous stirring in water at room temperature. The resulting precipitations were dried in vacuum and then heated to 450°C for 2 hr in air. **Figure 1** shows TEM images of the final product after thermal treatment, showing a layered V_2O_5 nanofiber with an average diameter of ~120 nm and a length in a scale of micrometres. The "rolling-up" formation of the layered nanofibers is subsequently proposed based on TEM observations, suggesting that [VO_6] building units were exfoliated into layered $K_{0.33}V_2O_5$ sheets at mild temperature with intercalated K-ions within between the layers with a large interplanar distance of 0.96 nm. The results suggested nanolayers rolled up along <001> direction, which appeared to be the most energetically favorable mechanism due to the large periodicity of the nanolayers.

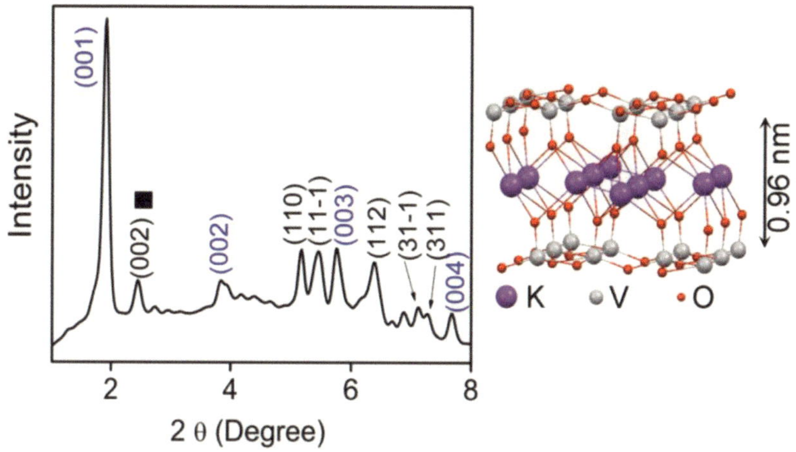

Figure 2. XRD patterns and crystalline schematics of $K_{0.33}V_2O_5$-layered nanofibers heat treated at 450°C. Wavelength of X-ray was 0.3196 nm. Diffractions from layered KxV_2O_5 (JCPDS 81045) and non-layered KxV_2O_5 (JCPDS 27718) were labeled in black and blue.

Figure 1c shows the high-resolution TEM (HRTEM) image of the final layered nanofiber with a uniform lamellar structure. An average repeating distance of 0.71 nm between the layers was observed over almost the entire layered nanofiber in the direction parallel to the longitudinal axis (growth direction). In addition to this periodic stacking of the layers, a clear fringe spacing of 0.33 nm was also observed from HRTEM imaging. **Figure 1d** shows the selected-area electron diffraction (SAED) pattern from the same layered nanofibers shown in **Figure 1c**, showing a single-crystalline nature of the layered nanofibers demonstrated by the discrete diffraction. The fringe distances were calculated from SAED to be 0.67 and 0.32 nm, matching well with the distances calculated from HRTEM. These values also matched well with d-spacing values from $K_{0.49}V_2O_5$ crystal (JCPDS 81045). The interplanar distance (d_{001} = 0.71 nm) obtained from HRTEM was smaller than that calculated from X-ray diffraction (XRD; 0.96 nm), which can be explained by the removal of structured water between the nanofibers' layers while under high vacuum and the consequent contraction of the interplanar distance of layered nanofibers during HRTEM analysis.

XRD patterns of the resulting product in which two series of reflections were shown in **Figure 2a**, which contained the conventional (hkl) set of reflection, and the (00l) set of diffraction with a high-degree ordering of the lamellar structures. The XRD reflection was in good agreement with potassium-intercalated layered V_2O_5 ($K_{0.49}V_2O_5$, JCPDS 81045), except that the peak at 2θ = 2.46° was indexed as the (002) plane of non-layered $K_XV_2O_5$ (JCPDS 27718). The strongest diffraction peak at 2θ = 1.98° was indexed to the (001) basal reflection with a corresponding d-spacing (d001) of 0.96 nm. Energy dispersive X-ray spectroscopy revealed the chemical composition of the final product that contained 83.7% V and 13.6% K (atomic percentage) with a chemical formula of $K_{0.33}V_2O_5$.

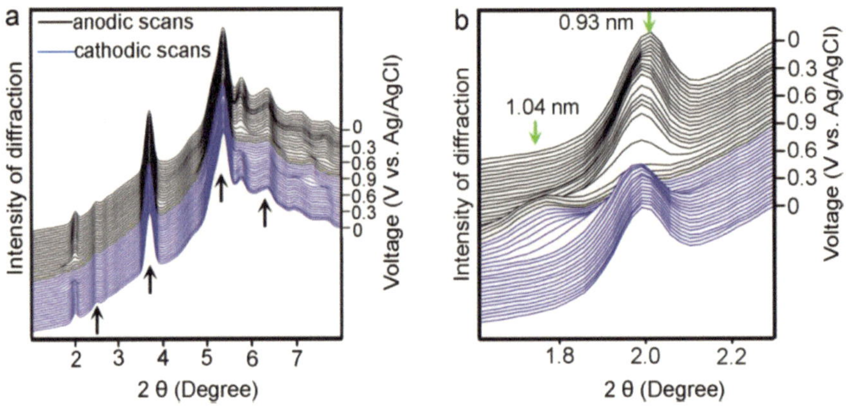

Figure 3. (a) *In situ* XRD of $K_{0.33}V_2O_5$-layered nanofibers at potentials from 0 to 0.9 V (vs. Ag/AgCl). The diffraction peaks from graphite oxide were marked by black arrows. (b) The evolution of the (001) planes upon the CV cycling.

Figure 3a shows a series of time-resolved XRD patterns of $K_{0.33}V_2O_5$-layered nanofibers using an electrochemical cell that the authors have reported in detail elsewhere. XRD patterns were obtained during anodic (black curves) and cathodic scans (blue curves) within the potential window between 0 and 0.9 V using a cyclic voltammetry (CV) program at a scan rate of 1 mV/s. Compared to the powder diffraction shown in **Figure 2**, steady diffraction background from the aqueous electrolyte solution and the commercial carbon paper from the in situ electrochemical cell were observed (marked by black arrows). The $K_{0.33}V_2O_5$-layered nanofibers showed varied peak intensities and positions at different potentials, suggesting dynamic changes of crystalline structure during the Faradaic processes, while the diffraction of layered nanofibers remained unchanged after one complete CV cycle. The results suggest that $K_{0.33}V_2O_5$-layered nanofibers possessed a reversible surface-controlled capacitive process, as well as a stable layered crystalline structure during the charge/discharge processes.

The changes of (001) diffraction peaks during the CV cycling are shown in **Figure 3b**. During the anodic scan, d-spacing of (001) plan (d_{001}) expanded from 0.93 nm ($2\theta = 1.98°$) to 1.04 nm ($2\theta = 1.76°$), corresponding to a 12% expansion of interplanar distance. On the other hand, d_{001} contracted from 1.04 to 0.95 nm during the cathodic scan. The expansion and contraction of d_{001} match very well with the charge storage mechanism of electrochemical extraction and insertion of K-ions into $K_{0.33}V_2O_5$-layered nanofibers. During the anodic scan, oxidation of $K_{0.33}V_2O_5$-layered nanofibers took place, accompanied with the removal of K-ions from interplanar layers of $K_{0.33}V_2O_5$. These processes can be expressed by the forward reaction of following equation:

$$K_{0.33}V_2O_5 \leftrightarrow K_{0.33-x}V_2O_5 + xK^+ + Xe^- \quad (5)$$

where x is the molar fraction of extracted K-ions. Depletion of K^+ weakened the electrostatic interaction between K-ions and negatively charged [VO_6] octahedral units within the layers, and increased the interplanar distance (d_{001}). Conversely, during the cathodic scans, $K_{0.33}V_2O_5$-layered nanofibers were reduced, accompanying with the insertion of K-ions into interplanar layers [backward reaction in Eq. (5)]. The strengthened electrostatic interaction between [VO_6] and K-ions decreased the interplanar distance.

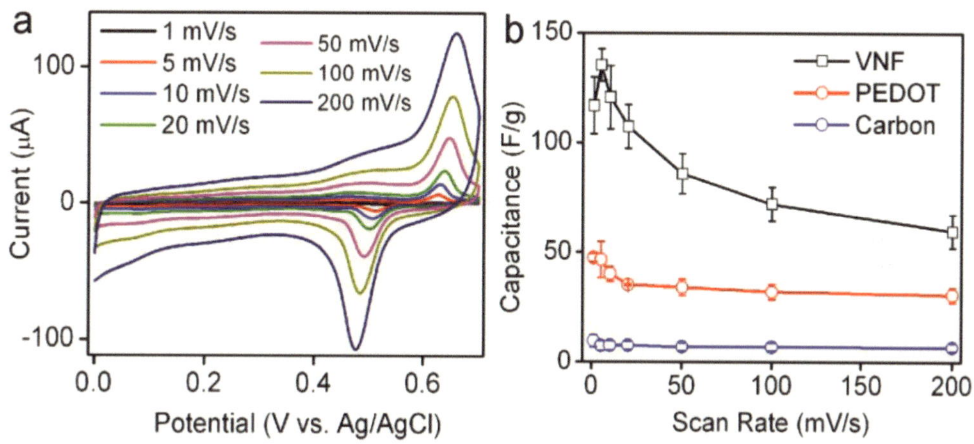

Figure 4. Three-electrode half-cell analyses: (a) CVs of $K_{0.33}V_2O_5$-layered nanofibers/PEDOT as loading of 5 and 1.25 μg, respectively; (b) gravimetric capacitance of $K_{0.33}V_2O_5$, PEDOT, and active carbon as functions of scan rates.

Figure 4a shows the cyclic voltammograms (CVs) of the mixture of $K_{0.33}V_2O_5$-layered nanofibers and PEDOT, respectively. The PEDOT and active carbon only electrodes showed broad redox peaks (data are not shown here), Indicating sluggish redox kinetics of K-ions. In contrast, the $K_{0.33}V_2O_5$/PEDOT electrode showed well-defined anodic and cathodic peaks (**Figure 4a**), representing the K^+ extraction and insertion behavior, occurring at ~0.6 (anodic scans) and ~0.5 V (cathodic scans), respectively. The calculated gravimetric capacitance (C_{MS}) was strongly dependent on the scan rate as shown in **Figure 4b**. As the scan rate decreased, C_{MS} increased. At a scan rate of 5 mV/s, the C_{MS} reached the maximum value of 136 F/g (based on $K_{0.33}V_2O_5$ mass). In general, the electrical conductivity of bulk V_2O_5 is low (10^{-5} ~ 10^{-3} Ω cm), so the addition of highly conductive PEDOT into $K_{0.33}V_2O_5$ improves the rate capability, although the C_{MS} value of the $K_{0.33}V_2O_5$/PEDOT electrode was associated strongly with the $K_{0.33}V_2O_5$ as shown in **Figure 4b**.

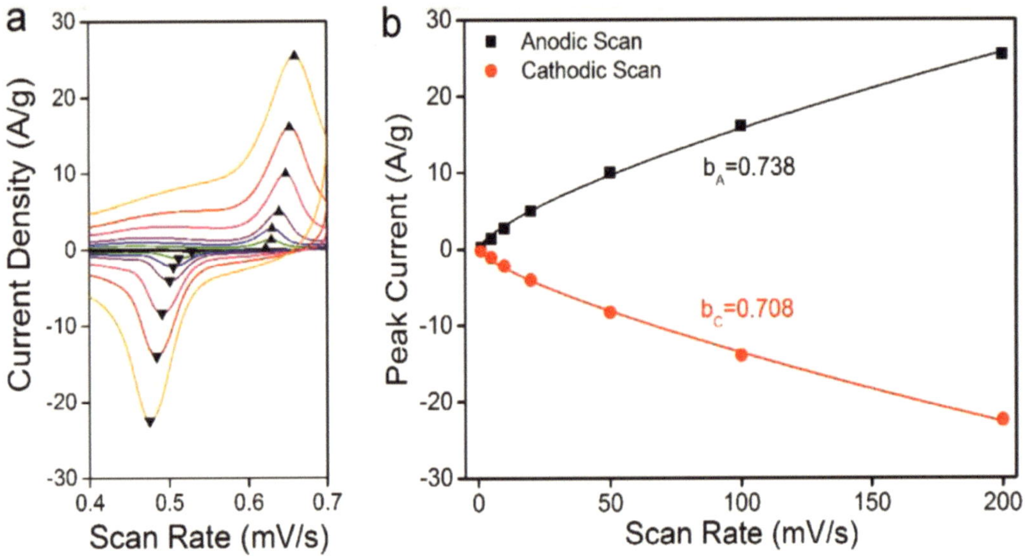

Figure 5. Calculated b-values of $K_{0.33}V_2O_5$-layered nanofibers from CV analyses at various potentials.

Charge-storage mechanism of K-ion within the $K_{0.33}V_2O_5$ electrode can be understood by analyzing current–voltage relationship provided by CV scans at different scan rates. Assuming that the peak current (i) obeyed the power law relationship with scan rate (v) at a given potential and was express as a combination of surface-controlled capacitive effects ($i_1 = k_1v$) and diffusion-controlled K-ion intercalation ($i_2 = k_2v^{1/2}$):

$$i = k_1 v + k_2 v^{1/2} \tag{6}$$

Accordingly, for the peak current response at a given potential can be described by a simple power law:

$$i = av^b \tag{7}$$

where a is adjustable parameter and b is equal to 0.5 or 1 when currents are strictly dominated by diffusion-controlled K-ion intercalation or surface-controlled capacitive effects. Thus, the contributions from capacitive charge and diffusion-limited redox charge during the CV cycling were extracted precisely. **Figure 5** shows typical power law regression analyses of i vs v plot for the both anodic and cathodic CV scans with scan rates ranging from 5 to 200 mv/s. The b-value was calculated to be 0.738 during the anodic scan and to be 0.708 during cathodic scan, indicating that combination of both surface-controlled capacitive process and diffusion-limited redox process contributed significantly to the redox behavior of $K_{0.33}V_2O_5$-layered nanofibers.

3. Highly disordered V_2O_5-layered electrode material for aqueous K-ion storage

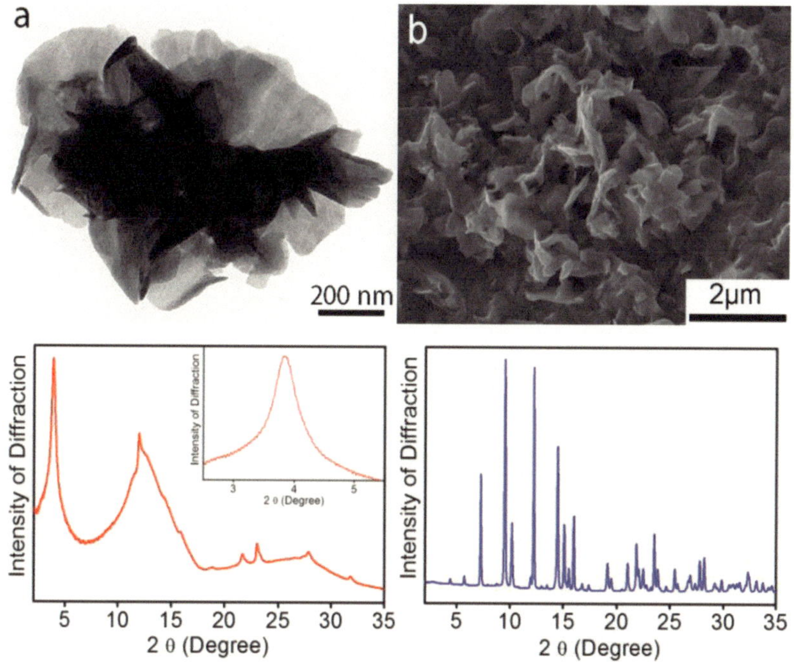

Figure 6. (a) TEM image of the $K_{0.2}V_2O_5$ nanosheets. (b) SEM image of network formed by the nanosheets. (c) X-ray diffraction ($\lambda = 0.72768$ Å) of the disordered nanosheets. (d) X-ray diffraction of bulk V_2O_5.

Disordered $K_{0.2}V_2O_5$ nanosheets were synthesized in an analogous way to the previously mentioned crystalline $K_{0.33}V_2O_5$ nanofibers, utilizing an aqueous reaction at room temperature between potassium hydroxide with vanadyl (IV) sulphate hydrate to form an amorphous VOx precursor. The amorphous VOx precursors then underwent a washing and drying process and were then thermally treated at low temperature and short time in air to form the disordered $K_{0.2}V_2O_5$ nanosheets. In **Figure 6a, b** TEM and SEM images of the disordered $K_{0.2}V_2O_5$ nanosheets are shown the sheet dimensions vary from sheet to sheet but are on the order of hundreds of nanometres. When loaded in more concentrated manner, they form a network of nanosheets. EDS measurements found that the atomistic ratio of the potassium to vanadium ratio was 9.731% K to 9.0269% V, resulting in the chemical formula of $K_{0.2}V_2O_5$.

The XRD spectra of the disordered nanosheets in **Figure 6c** contained only a few very broad peaks due to the large degree of disorder within the structure. The strong low angle peak at 3.82° 2θ was indexed as the (001) plane. The d-spacing between the layers of the disordered nanosheets is 10.916, but other structural data are hard to extract due to the severe broadening caused by the disorder. In contrast, **Figure 6d** is the XRD spectra of highly crystalline bulk V_2O_5. It contains very sharp well-defined peaks and can be easily identified as orthorhombic structure with Pmmn symmetry and lattice parameters of a = 3.563, b = 11.50870, c = 4.375 (**Figure 7**).

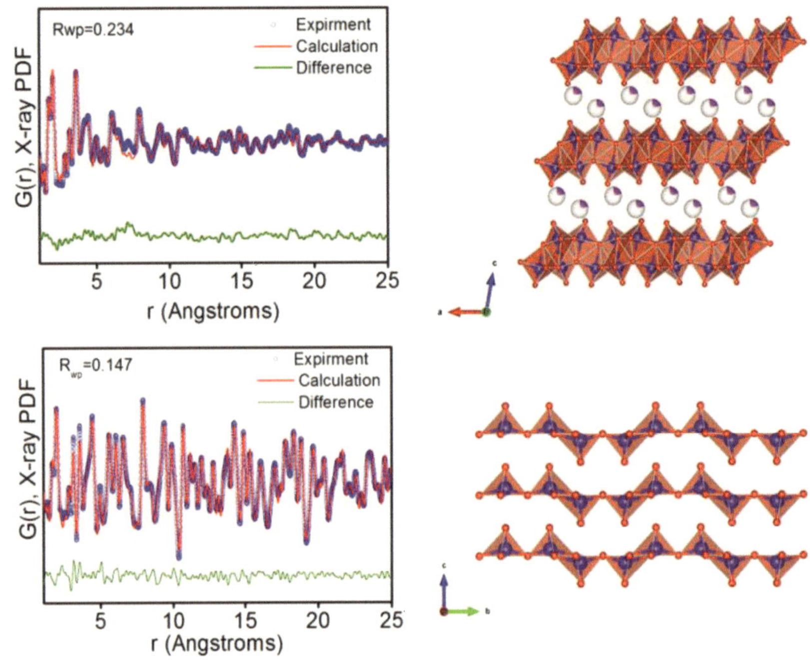

Figure 7. X-ray atomic pair distribution function analysis: (top left) disordered nanosheets and (top right) the solved crystal structure of bi-layered $[VO_6]$ octahedral layers with a potassium intercalate. (Bottom left) bulk V_2O_5 and (bottom right) the bulk crystal structure.

Due to lack of crystallinity, there is much broadening of the peaks in the XRD spectra, and the atomistic crystal structure was not able to be accurately identified. Therefore, the local structure of the disordered nanosheets was characterized by total X-ray total scattering experiments and refinement of the atomic pair distribution function (PDF). The PDF is calculated by taking the sine Fourier transform of the total structure factor and is a real-space representation of inter atomic distances through a radial distribution function. PDF takes into account both the Bragg diffraction and diffuse scattering caused by local nonperiodic structural defects allowing for the investigation of disordered materials. Total scattering experiments can be done with both neutrons and X-rays allowing for good resolution between elements.

The major phase identified in the PDF of the disordered nanosheets was a monoclinic structure with a C2/m symmetry. It was made of V_2O_5 bilayers consisting of distorted $[VO_6]$ octahedra and a potassium intercalate within the bilayers. The bulk orthorhombic V_2O_5 phase consisting of $[VO_5]$ polyhedra was also detectable in the PDF of the disordered nanosheets. Based in the refinement, the sample was comprised of 72% monoclinic phase and 28% of the orthorhombic phase by weight. The short range order in the disordered nanosheets is dominated by the monoclinic phase, which is only ordered out to 14 Å, accounting for the initial sharp decrease

http://dx.doi.org/10.5772/62759

in the intensity of the PDF. The long range order is dominated by orthorhombic phase. The X-ray PDF of the bulk V_2O_5 is sharper and more intense as compared to the nanosheets due to the high crystallinity. The results of the real-space refinement of the PDF and the Rietveld refinement in reciprocal space of the XRD spectra are nearly identical.

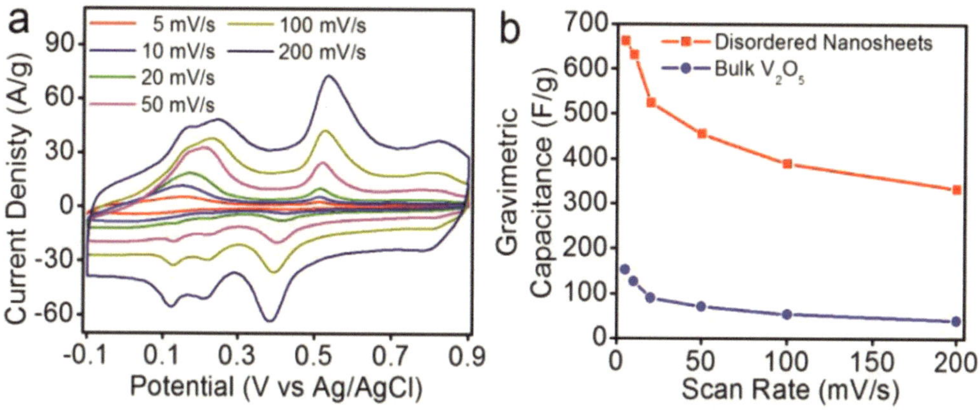

Figure 8. Three-electrode half-cell analyses: (a) CVs of $K_{0.2}V_2O_5$ disordered nanosheets/PEDOT as loading of 5 and 1.25 μg, respectively. (b) Gravimetric capacitance of $K_{0.2}V_2O_5$, bulk V_2O_5 and as a function of scan rate.

The electrochemical behavior of the layered $K_{0.2}V_2O_5$ disordered nanosheets material was investigated utilizing a three-electrode half-cell with a rotating disk electrode. Aqueous inks were made with the 4:1 ratio of disordered nanosheets material to poly (3,4-ethylenedioxy-thiophene) (PEDOT), a conductive polymer. CV measurements were conducted in a neutral aqueous 1 M KCl electrolyte solution with a 1 V potential window from −0.1 to 0.9 V versus a saturated KCl Ag/AgCl reference electrode. CV measurements at various scan rates between 5 and 200 mV/s are shown in **Figure 8a**. Three distinct oxidation peaks are observable in the anodic scan at 0.20 V, 0.53 V, and 0.78 V. The corresponding peaks of all three redox couples can be observed in the cathodic scan at 0.16 V, 0.38 V, and 0.79 V. All of the redox peaks remain visible throughout all of the scan rates tested.

The $K_{0.2}V_2O_5$ nanosheets exhibit excellent capacitance at low scan rates, 661 F/g (183 mAh/g) at a scan rate of 5 mV/s in 1 V potential window. A significant amount of capacitance was retained even at higher scan rates, and 334 F/g (93 mAh/g) at 200 mV/s. 52% of the specific capacitance was retained with a 40-times increase in scan rate, from 5 to 200 mV/s. Assuming one charge transfer per vanadium atom, the theoretical capacitance of V_2O_5 is 294 mAh/g, 0.6 electron transfer per vanadium atom. As shown in Eq. (4), the amount of potassium atoms inserted is directly proportional to the charge stored, when fully reduced at 5 mV/s the amount of potassium inserted is equivalent to $K_{1.2}V_2O_5$ and at 200 mV/s the amount of potassium inserted is equivalent to $K_{0.6}V_2O_5$.

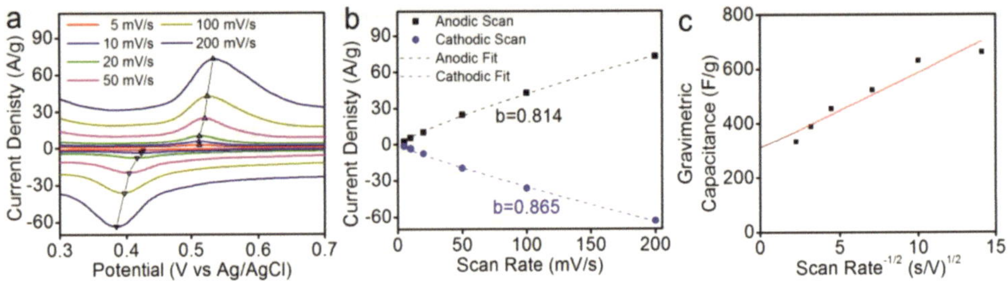

Figure 9. (a) Anodic and cathodic redox peaks from the CV measurement of the disordered $K_{0.2}V_2O_5$ nanosheets at 0.53 and 0.38 V, respectively, and (b) the corresponding b-value calculation. (c) Infinite sweep rate extrapolation.

To further investigate the charge storage mechanism, the data from CV measurements was further exploited with electrokinetic analysis. The scan rate (v) vs. peak current for the redox couple shown in **Figure 9a** was analyzed with the b-value evaluation from Eq. (6). The results of the fitting are shown in **Figure 9c**. The anodic scan had a b-value of 0.814, and the cathodic scan had a b-value of 0.865. Both processes seem to be similar in mechanism having similar b-values, with the oxidation process associated with the anodic scan being slightly more diffusion limited than the reduction process associated with the cathodic scan. As compared to the crystalline $K_{0.33}V_2O_5$ nanofibers, the b-value of the disordered nanosheets is closer to 1; evidence that the disordered nanosheets are less diffusion limited then crystalline nanofibers.

Further information can be obtained by extrapolation to an infinite scan rate. The total charge (q_T) stored can be expressed as the sum of the charge stored at the surface by capacitive charge storage process and the charge stored in the bulk by diffusion limited diffusion process. The charge stored on the surface is from capacitive processes, that is, double-layer/under potential deposition, charge separation at the surface, and pseudocapacitive reversible near-surface redox reactions and is assumed to be constant (q_∞) with varying scan rate. Semi-infinite linear diffusion is assumed for the intercalation process into the bulk and is approximated with a constant (c) multiplied by the inverse square root of scan rate ($v^{-1/2}$)

$$q_T(v) = q_\infty + cv^{-1/2} \tag{8}$$

The extrapolated specific capacitance for the disordered $K_{0.2}V_2O_5$ nanosheets at an infinite scan rate is 308 F/g. Thus, meaning that capacitive processes account for 308 F/g of the charge capacitance. At a scan rate of 5 mV/s, the capacitive processes account for 46% of the capacitance and at 200 mV/s, the capacitive processes account for 92% of the capacitance.

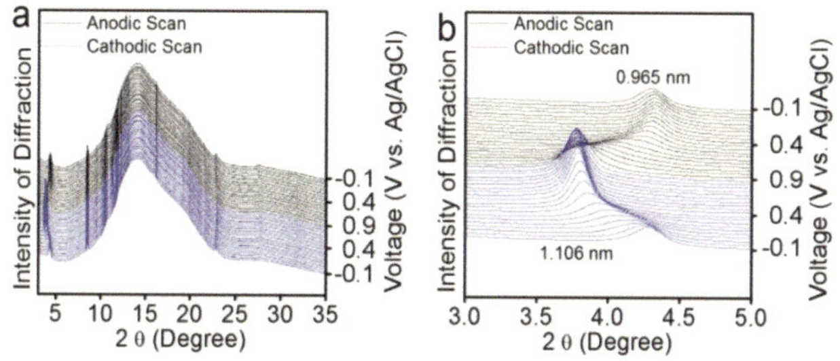

Figure 10. (a) *In situ* XRD ($\lambda = 0.72768$ Å) of the $K_{0.2}V_2O_5$ disordered nanosheets at potentials from −0.1 to 0.9 V (vs. Ag/AgCl). (b) The evolution of the (001) plane upon the CV cycling.

The intercalation of potassium ions was studied through *in situ* XRD measurement while electrochemical cycling. In **Figure 10a**, the full XRD spectra for all of the scans collected during the second cycle of the CV measurement are shown. During the anodic scan the (001), peak shifts to lower angle continuously stating an applied potential of 0.4 V all the way to 0.9 V. The d-spacing increased from 0.965 to 1.106 nm. The removal of potassium causes the oxidation of the vanadium from V^{4+} to the V^{5+} increasing the van der Waals repulsion between bilayers, which are made of the negatively charged $[VO_6]^-$ octahedra units, in turn causing the expansion in the interlayer distance.

Conversely, during the cathodic scan, the (001) peak shifts back to higher 2θ angle, decreasing the d_{001} spacing from 1.106 to 0.965 nm. The shift (001) peak during the cathodic scan is also continuously occurring between 0.4 and −0.1 V. The insertion of potassium causes the reduction of vanadium from V^{5+} to V^{4+} and causes the van der Waals repulsion between bilayers to decrease, and therefore, the d_{001} spacing decreases. The (001) peak returns to the starting 2θ angle, suggesting that the intercalation/de-intercalation of the K^+ ions are reversible in nature.

The (020) peak also shifts during cycling but at an order of magnitude smaller than that of the (001) peak. It also shifted in the opposite direction with the (001) layered peak, increasing the d_{020} upon reduction, and decreasing d_{020} upon oxidation. This shifting is caused by the increase and decrease of the V–O bond distance. The higher oxidation state of vanadium resulted in a shorter V–O bond. As with the (001) peak, the shifting in (020) appears to be completely reversible.

4. Summary

The emergence of in situ time-resolved techniques, including X-ray, neutron, and electron miscopy tools, provides us with an unprecedented opportunity to "observe," in nanoscale at millisecond time steps, the electrochemical processes in a close-to-real operation conditions. At the same time, the advent of high-performance computational tools and advance computational science methods are allowing us to analyze and predict electrochemical reactions at the nanoscale and use big data tool sets to "see" how an electrochemical reaction drives functionality under real operation conditions. In this study, we have shown the power of combining advanced experimental, analytical, and computational tools, which help move beyond current understanding of electrochemistry and reaction engineering, in case of highly crystalline and highly disordered V_2O_5-layered nanostructure, and help leap forward into complex reaction kinetics to find new ways to convert and store energy.

Acknowledgements

This work was supported by the US Department of Energy (DOE), Office of Science, Basic Energy Sciences under Award # DE-SC0010286. Use of the National Synchrotron Light Source, Brookhaven National Laboratory, was supported by the US DOE, Office of Science, Office of Basic Energy Sciences, under Contract No. DE-AC02-98CH10886. This research used resources of the Advanced Photon Source, a US DOE Office of Science User Facility operated for the DOE Office of Science by Argonne National Laboratory under Contract No. DE-AC02-06CH11357. Research conducted at ORNL's Spallation Neutron Source was sponsored by the Scientific User Facilities Division, Office of Basic Energy Sciences, US DOE.

Author details

Daniel S. Charles and Xiaowei Teng*

*Address all correspondence to: xw.teng@unh.edu

Department of Chemical Engineering, University of New Hampshire, Durham, USA

References

[1] Hoffert MI, Caldeira K, Benford G, Criswell DR, Green C, Herzog H, et al. Advanced technology paths to global climate stability: Energy for a greenhouse planet. Science. 2002;298:981–987. doi:10.1126/science.1072357

[2] Turner JA. A realizable renewable energy future. Science. 1999;285:687–689. doi: 10.1126/science.285.5428.687

[3] Chiang YM. Building a better battery. Science. 2010;330:1485–1486. doi:10.1126/science.1198591

[4] Palomares V, Casas-Cabanas M, Castillo-Martinez E, Han MH, Rojo T. Update on Na-based battery materials. A growing research path. Energy & Environmental Science. 2013;6:2312–2337. doi:10.1039/c3ee41031e

[5] Li Z, Young D, Xiang K, Carter WC, Chiang YM. Towards high power high energy aqueous sodium-ion batteries: The $NaTi_2(PO_4)_3/Na_{0.44}MnO_2$ system. Advanced Energy Materials. 2013;3:290–294. doi:10.1002/aenm.201200598

[6] Stevens DA, Dahn JR. High capacity anode materials for rechargeable sodium-ion batteries. Journal of the Electrochemical Society. 2000;147:1271–1273. doi: 10.1149/1.1393348

[7] Suo L, Borodin O, Gao T, Olguin M, Ho J, Fan X, et al. "Water-in-salt" electrolyte enables high-voltage aqueous lithium-ion chemistries. Science. 2015;350:938–943. doi:10.1126/science.aab1595

[8] Gleiter H. Diffusion in nanostructured metals. Physica Status Solidi B-Basic Research. 1992;172:41–51. doi:10.1002/pssb.2221720106

[9] Athouel L, Moser F, Dugas R, Crosnier O, Belanger D, Brousse T. Variation of the MnO_2 birnessite structure upon charge/discharge in an electrochemical supercapacitor electrode in aqueous Na_2SO_4 electrolyte. Journal of Physical Chemistry C. 2008;112:7270–7277. doi:10.1021/jp0773029

[10] Fukuda K, Nakai I, Ebina Y, Ma RZ, Sasaki T. Colloidal unilamellar layers of tantalum oxide with open channels. Inorganic Chemistry. 2007;46:4787–4789. doi:10.1021/ic7004002

[11] Ida S, Ogata C, Unal U, Izawa K, Inoue T, Altuntasoglu O, et al. Preparation of a blue luminescent nanosheet derived from layered perovskite $Bi_2SrTa_2O_9$. Journal of the American Chemical Society. 2007;129:8956–8957. doi:10.1021/ja073105b

[12] Kai K, Yoshida Y, Kageyama H, Saito G, Ishigaki T, Furukawa Y, et al. Room-temperature synthesis of manganese oxide monosheets. Journal of the American Chemical Society. 2008;130:15938–15943. doi:10.1021/ja804503f

[13] Liu Z, Ma R, Ebina Y, Takada K, Sasaki T. Synthesis and delamination of layered manganese oxide nanobelts. Chemistry of Materials. 2007;19:6504–6512. doi:10.1021/cm7019203

[14] Song MS, Lee KM, Lee YR, Kim IY, Kim TW, Gunjakar JL, et al. Porously assembled 2D nanosheets of alkali metal manganese oxides with highly reversible pseudocapacitance behaviors. Journal of Physical Chemistry C. 2010;114:22134–22140.

[15] Yeager M, Du WX, Si R, Su D, Marinkovic N, Teng XW. Highly efficient $K_{0.15}MnO_2$ birnessite nanosheets for stable pseudocapacitive cathodes. Journal of Physical Chemistry C. 2012;116:20173–20181. doi:10.1021/jp304809r

[16] Mas-Balleste R, Gomez-Navarro C, Gomez-Herrero J, Zamora F. 2D materials: To graphene and beyond. Nanoscale. 2011;3:20–30. doi:10.1039/C0NR00323A

[17] Tepavcevic S, Xiong H, Stamenkovic VR, Zuo XB, Balasubramanian M, Prakapenka VB, et al. Nanostructured bilayered vanadium oxide electrodes for rechargeable sodium-ion batteries. ACS Nano. 2012;6:530–538. doi:10.1021/nn203869a

[18] Yeager MP, Du WX, Bishop B, Sullivan M, Xu W, Su D, et al. Potassium-ion storage in layered $K_{0.33}V_2O_5$ electrodes for aqueous pseudocapacitors. ChemSusChem. 2013;6:2231–2235. doi:10.1002/cssc.201300480